Contents

Volume 1
Solutions to Selected Exercises

EXPERIMENTAL RESEARCH DESIGN and ANALYSIS

A Practical Approach for Agricultural and Natural Sciences

A. Reza Hoshmand

CRC Press

Boca Raton   Ann Arbor   London   Tokyo

Catalog record is available from the Library of Congress
ISBN: 0-8493-9468-6

Chapter 2

KEY ASSUMPTIONS OF EXPERIMENTAL DESIGNS

EXERCISES

1. Using the data given below, perform Tukey's test to determine if the data meet the assumption of additivity.

Solution:

Treatment	Block				$\overline{X}_{i.}$	$d_i = \overline{X}_{i.} - \overline{X}_{..}$	$w_i = \sum X_{ij} d_j$
	I	II	III	IV			
A	20	32	22	24	24.5	0.6875	15.875
B	18	20	23	25	21.5	-2.3125	20.875
C	19	21	28	27	23.8	-0.0625	26.8125
D	24	24	26	28	25.5	1.6875	10.625
Mean $(\overline{X}_{.j})$	20.25	24.25	24.75	26			
$d_j = (\overline{X}_{.j} - \overline{X}_{..})$	-3.563	0.438	0.938	2.188			
$\overline{X}_{..}$				23.81			

To be able to perform the Tukey's test, we must first perform the ANOVA test on the data as shown below. See chapter 4 on how the analysis is performed.

Treatment	Block				Tr. Total	Treatment Mean
	I	II	III	IV		
A	20	32	22	24	98	24.5
B	18	20	23	25	86	21.5
C	19	21	28	27	95	23.75
D	24	24	26	28	102	25.5
Grand Total					381	
Grand Mean						23.8125

Correction Factor = 9072.6

Degree of Freedom = d.f.
Total d.f. = 16 - 1 = 15
Treatment d.f. = 4 - 1 = 3
Error d.f. = 12

Based on the analysis of the data, the treatment, block, and error mean squares are shown below:

<u>ANOVA of the Data Presented</u>

Source of Variation	Degrees of Freedom	Sum of Squares	Mean Square	F
Block	3	74.19	24.73	2.068*ns*
Treatment	3	34.69	11.56	0.967*ns*
Error	9	107.60	11.96	
Total	15	216.48		

ns = nonsignificant.

Now that we have performed the ANOVA test, we will partition the error sum of squares to its components and determine if the additivity assumption holds.

Now we are ready to perform the Tukey's test.

$$w_i = \sum X_{ij} d_j$$

$$N = \sum w_i d_i = \sum \sum X_{ij} d_i d_j = -21.1$$

$$\sum d_i^2 = 8.6719$$

$$\sum d_j^2 = 18.547$$

$$SS \text{ nonadditivity} = \frac{(-21.1)^2}{(8.6719)(18.547)} = 2.769$$

ANOVA Table and the Test of Additivity

Source of Variation	Degrees of Freedom	Sum of Squares	Mean Square	F
Error (Block x Treatment)	9	107.60	11.96	
Nonadditivity	1	2.77	2.77	<1
Residual	8	104.83	13.10	

The result indicates a strong evidence that the assumption of additivity is correct.

2. To determine the impact of 2 newly developed fungicides on roses, a horticulturist used a randomized block experiment. Each block contained 6 plots, and each fungicide was used on 2 of the plots within each block. Two weeks after the spray, 50 leaves were examined on each plot, and the number of leaves showing rust problems were recorded as shown below. Determine whether any transformation is required before an analysis could be performed.

	Fungicide					
Block	F1		F2		Control	
I	6	8	7	9	25	18
II	5	6	8	5	18	19
III	3	4	7	8	14	26

Solution:

To determine whether any transformation is required, we are basically interested in finding out whether the data meet the *normality* assumption. We have to look at the error terms associated with each observation. To find the error term for any cell of a 2-way table, we use the following equation:

$$e_{ij} = X_{ij} - \mu - t_i - b_j$$

The treatment and block effect which are used in the above equation are found:

$$\text{Treatment effect } (t_i) = \overline{X}_{i.} - \mu$$

$$\text{Block effect } (b_j) = \overline{X}_{.j} - \mu$$

We calculate the mean and each of the error terms as shown below:

Block	FI-A	FI-B	FII-A	FII-B	FIII-A	FIII-B	Mean $\overline{X}_i$
I	6.0	8.0	7.0	9.0	25.0	18.0	12.2
II	5.0	6.0	8.0	5.0	18.0	19.0	10.2
III	3.0	4.0	7.0	8.0	14.0	26.0	10.3
Mean ($\overline{X}_{.j}$)	4.7	6.0	7.3	7.3	19.0	21.0	
$\overline{X}_{..} = \mu$							10.9

(Treatment)

Each error term is computed by substituting the treatment and the block effect.

$$e_{ij} = X_{ij} - \mu - (\overline{X}_{i.} - \mu) - (\overline{X}_{.j} - \mu)$$

Block	F1-A	F1-B	F2-A	F2B	F3-A	F3-B	Total
I	0.06	0.72	-1.61	0.39	4.72	-4.28	0.00
II	1.06	0.72	1.39	-1.61	-0.30	-1.28	0.00
III	-1.11	-1.40	0.22	1.22	-4.40	5.56	0.00
IV	0.00	0.00	0.00	0.00	0.00	0.00	0.00
Total	0.00	0.00	0.00	0.00	0.00	0.00	

(Fungicide)

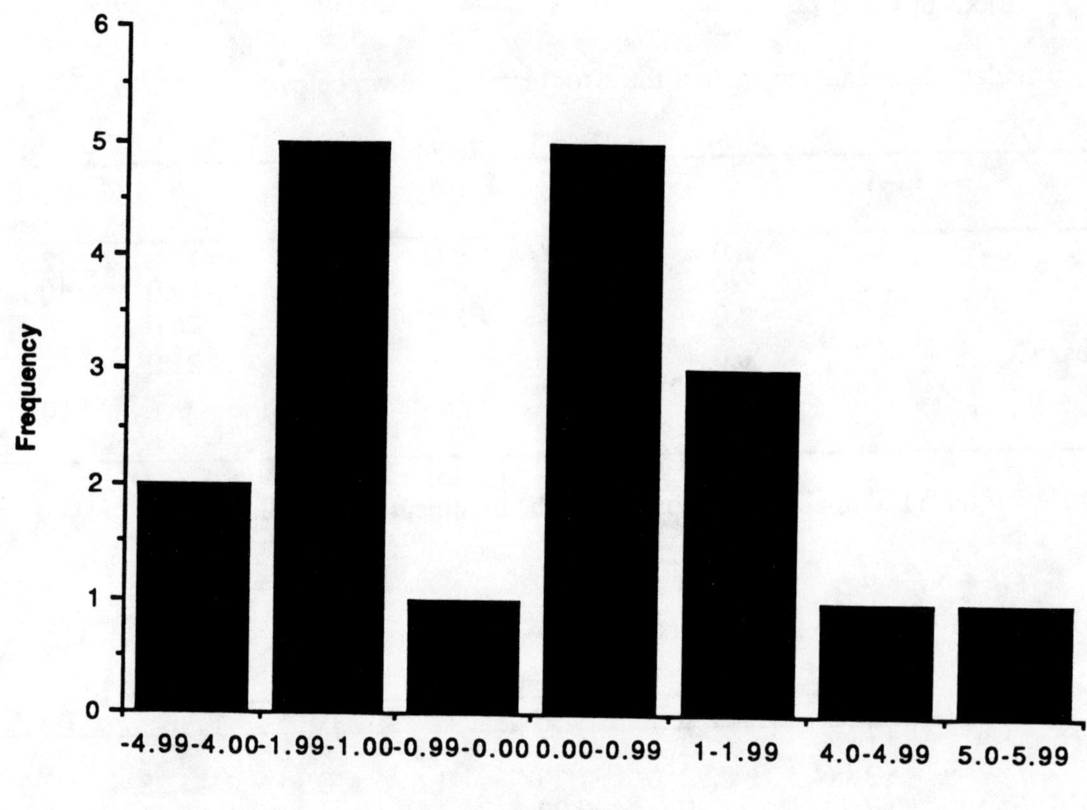

Looking at the error terms, we note that the distribution does not appear to be normal, as there are two modal classes. Thus, the normality assumption is untenable. We must transform the data (logarithmic transformation) before any analysis is carried out.

3. Using a randomized complete block design, an animal scientist performed an experiment in which the impact of linseed cake meal on weight gain (in pounds) of several groups of animals were observed. Perform a logarithmic transformation and analysis on the data.

Treatment (Species)	Block			
	I	II	III	IV
Pig- Linseed Cake	95	102	110	96
Pig- Control	82	85	80	75
Sheep-Linseed Cake	112	130	114	129
Sheep-Control	90	86	89	85
Chicken-Linseed Cake	3.0	2.9	2.6	2.8
Chicken-Control	1.5	1.9	2.0	1.2

Solution:

<u>Log (x +1) of the data</u>

				Mean	Total
1.9822	2.0128	2.0453	1.9858	8.0261	2.0065
1.9191	1.9345	1.9085	1.8808	7.6429	1.9107
2.0531	2.1173	2.0607	2.1139	8.3450	2.0863
1.9590	1.9395	1.9542	1.9345	7.7872	1.9468
0.6021	0.5911	0.5563	0.5798	2.3293	0.5823
0.3979	0.4624	0.4771	0.3424	1.6798	0.4200

After we have transformed the data, there seems to be a much more consistent pattern about the transformed data, though there is still some random variation.

We now perform the Bartlett's test of homogeneity of variance to see if the transformed data meet this assumption.

Treatment	Mean	s_i^2	Coded s_i^2	Log Coded s_i^2
Pig- Linseed Cake	2.0065	0.00085	8.547	2.1456
Pig- Control	1.9107	0.00051	5.119	1.6331
Sheep-Linseed Cake	2.0863	0.00116	11.601	2.4511
Sheep-Control	1.9468	0.00014	1.361	0.3079
Chicken-Linseed Cake	0.5823	0.00038	3.839	1.3452
Chicken-Control	0.4200	0.00305	30.467	3.4166
Total	8.9526		60.934	11.2995
Mean			10.155	
Log of Mean			2.318	

Since the value of the variance for each treatment is less than one, we have the problem of negative logarithms. For this reason each variance was multiplied by a constant (in the present case 10,000) to remedy the problem. The coded variance (Coded s_i^2) is the original variance multiplied by 10,000.

The Bartlett's test of homogeneity of variance is:

$$\chi^2 = \frac{M}{C}$$

$M = (3)[6(2.318)-11.2995]$

$M = 7.838$

$C = 1 + (6+1)/(3)(6)(3) = 1.13$

$$\chi^2 = \frac{7.838}{1.13}$$
$$\chi^2 = 6.94$$

Since the Table chi square for the 5 degrees of freedom (1 less than the number of treatments) is 11.07 at 5%, thus we cannot reject the null hypothesis that the variance are homogeneous.

4. In a randomized experiment with 3 replications, the number of insects found per plot after the fields were sprayed with different rates of a general insecticide. Verify the homogeneity of variance for the data using Bartlett's test.

Number of insects found per plot following different insecticide treatments

Treatment	Rep. I	Rep. II	Rep. III
A	10	12	7
B	9	15	10
C	5	16	12
D (Control)	15	22	19

Solution:

Log (x +1) of the data

Rep. I	Rep. II	Rep. III	Total	Mean
1.0414	1.1139	0.9031	3.0584	1.0195
1.0000	1.2041	1.0414	3.2455	1.0818
0.7782	1.2304	1.1139	3.1225	1.0408
1.2041	1.3617	1.3010	3.8668	1.2889

Treatment	Mean	s_i^2	Coded s_i^2	Log Coded s_i^2
A	1.0195	0.0115	11.50	2.4397
B	1.0818	0.0116	11.60	2.4510
C	1.0408	0.0551	55.10	4.0091
D (Control)	1.2889	0.0063	6.30	1.8405
Total	4.4310		84.50	10.7403
Mean			21.13	
Log of Mean			3.05	

Since the value of the variance for each treatment is less than one, we have the problem of negative logarithms. For this reason each variance was multiplied by a constant (in the present case 1000) to remedy the problem. The coded variance (Coded s_i^2) is the original variance multiplied by 1000.

The Bartlett's test of homogeneity of variance is:

$$\chi^2 = \frac{M}{C}$$

$M = (2)[4(3.05)-10.7403]$

$M = 2.919$

$C = 1 + (4+1)/(3)(4)(2) = 1.21$

$$\chi^2 = \frac{2.919}{1.21}$$

$$\chi^2 = 2.41$$

Since the Table chi square for the 3 degrees of freedom (1 less than the number of treatments) is 7.815 at 5%, thus we cannot reject the null hypothesis that the variance are homogeneous.

5. Slow germination limits establishment of perennial warm season forage grasses, and temperature is a major factor influencing it. A crop scientist conducted a completely randomized design experiment in which the following percentages of seeds germinated were recorded.

Influence of temperature on germination percentage of seeds of warm-season forage grasses

Species	Temperature, °C					
	9	12	15	20	25	30
Native big blue stem	6	32	36	30	27	20
Rountree big blue stem	36	58	68	64	59	49
Caucasian bluestem	0	1	8	10	12	14
Blackwell switch grass	4	14	18	8	9	12
Osage	30	52	56	54	44	32

(a) What type of transformation should be used?
(b) Perform an analysis of variance on the data.

Solution:

a. Since this involves a percentage rate of germination, the Arc Sine is the appropriate transformation. The transformed data are given below:

Transformed Data using the Arc Sine Transformation

							Total
Native big blue stem	14.18	34.45	36.87	33.21	31.31	26.56	176.6
Rountree big blue stem	36.87	49.60	55.55	53.13	50.18	44.43	289.8
Caucasian bluestem	0.00	5.74	16.43	18.44	20.27	21.97	82.9
Blackwell switch grass	11.54	21.97	25.10	16.43	17.46	20.27	112.8
Osage	33.21	46.15	48.45	47.29	41.55	34.45	251.1
Total	95.80	157.91	182.40	169.00	161.00	147.68	913.1

The correction factor is:

$$C = \frac{G^2}{ab} = \frac{(913.1)^2}{30} = 27789.3$$

Total Sum of Square = 6464.54

Sum of square for seed variety = 5173.01

Sum of square for temperature = 897.033

Error sum of square = 394.503

ANOVA Table for Seed Varieties and Temperature

Source of Variation	Degrees of Freedom	Sum of Squares	Mean Square	F
Seed Variety	4	5173.01	1293.25	65.56**
Temperature	5	897.03	179.41	9.09**
Error	20	394.50	19.73	
Total	29	6464.54		

The results show that both the seed variety and temperature are highly significant.

Chapter 4

SINGLE-FACTOR EXPERIMENTAL DESIGNS

1. Field measurements were made to study the response of field-grown cassava (*Manihot esculenta* Crantz) to changes in the application of a fertilizer. The researcher conducted a completely randomized design to find out if there were any differences in the amount of the dry matter produced under 5 different fertilizer regimes. The following data were collected from the experiment with 4 replications:

Dry matter production (tons/ha) of cassava as a result of 5 different fertilizer applications

Treatment	Rep. I	Rep. II	Rep. III	Rep. IV
Control	2.20	2.10	2.25	2.01
50 kgs/ha	2.40	2.56	2.66	2.52
100 kgs/ha	2.60	2.68	2.79	2.66
150 kg/ha	3.00	3.56	4.00	4.66
200 kg/ha	3.50	4.98	5.00	4.20

(a)   Perform the analysis of variance.
(b)   Are there differences between the treatments?
(c)   Compute the coefficient of variation? What does this value mean?

Solution:

a.

Treatment	Rep. I	Rep. II	Rep. III	Rep. IV	Treatment Total	Tr. Mean
Control	2.2	2.10	2.25	2.01	8.56	2.140
50 kgs/ha	2.4	2.56	2.66	2.52	10.14	2.535
100 kgs/ha	2.6	2.68	2.79	2.66	10.73	2.683
150 kg/ha	3.0	3.56	4.00	4.66	15.22	3.805
200 kg/ha	3.5	4.98	5.00	4.20	17.68	4.420
Grand Total						62.33
Grand Mean						3.12

$$C = \frac{G^2}{N} = \frac{(62.33)^2}{20} = 194.3$$

$$\text{Total } SS = \sum X_i^2 - C$$

$$\text{Total } SS = [(2.2)^2 + (2.4)^2 + \ldots\ldots + (4.6)^2 + (4.2)^2] - 194.3$$

$$= 17.72$$

$$\text{Treatment } SS = \frac{\sum T_i^2}{r} - C$$

$$\text{Treatment } SS = \frac{(8.56)^2 + (10.14)^2 + \ldots(17.68)^2}{4} - 194.3$$

$$\text{Treatment } SS = 14.61$$

Error SS = Total SS - Treatment SS

$$= 17.72 - 14.61$$

$$= 3.11$$

Analysis of Variance of Dry Matter Production of Cassava from 5 Different Fertilizer Applications

Source of Variation	Degree of Freedom	Sum of Squares	Mean Square	F
Treatment	4	14.61	3.65	12.92**
Error	11	3.11	0.28	
Total	15	17.72		

** = Highly significant or significant at 1% level.

b. Since the computed value of F is greater than the table value of 5.67 at the 1% level of significance, we conclude that there is a significant difference between the treatments.

c. The coefficient of variation is computed as:

$$cv = \frac{\sqrt{\text{Error } MS}}{\text{Grand mean}} \times 100$$

$$cv = 17.07$$

The cv indicates the degree of precision with which the treatments are compared. In actuality, cv expresses the experimental error as a percentage of the mean. Thus, the higher the value of the cv, the lower is the reliability of the experiment.

2. Assume that the data collected in Exercise 1 has some observations missing, as shown below:

Dry matter production (tons/ha) of cassava as a result of 5 different fertilizer applications

Treatment	Rep. I	Rep. II	Rep. III	Rep. IV
Control	2.20	2.10	2.25	2.01
50 kgs/ha	2.40	----	2.66	2.52
100 kgs/ha	2.60	2.68	----	2.66
150 kg/ha	----	3.56	4.00	4.66
200 kg/ha	3.50	4.98	5.00	4.20

(a) Perform an analysis of variance on the data.
(b) What can you say about the results of this exercise as compared to Exercise 1?
(c) Compute the coefficient of variation.
(d) Is there an difference between the cv computed here and in exercise 1? What could you say about the differences?

Solution:

a.

Treatment	Rep. I	Rep. II	Rep. III	Rep. IV	Treatment Total	Treatment Mean
Control	2.2	2.1	2.25	2.01	8.56	2.14
50 kgs/ha	2.4	0.0	2.66	2.52	7.58	2.53
100 kgs/ha	2.6	2.68	0.0	2.66	7.94	2.65
150 kg/ha	0.0	3.56	4.0	4.66	12.22	4.07
200 kg/ha	3.5	4.98	5.0	4.20	17.68	4.42
Grand Total						53.98
Grand Mean						3.18

The correction factor is:

$$C = \frac{G^2}{N} = \frac{(53.98)^2}{20} = 145.69$$

$$\text{Total } SS = \sum X_i^2 - C$$

$$\text{Total } SS = [(2.2)^2 + (2.4)^2 + \ldots + (4.6)^2 + (4.2)^2] - 145.69$$

$$= 42.94$$

$$\text{Treatment } SS = \sum \frac{T_i^2}{r_i} - C$$

$$= \left[\frac{(8.56)^2}{4} + \frac{(7.58)^2}{3} + \frac{(7.94)^2}{3} + \ldots + \frac{(17.68)^2}{4} \right] - 145.69$$

$$= 40.72$$

$$\text{Error } SS = \text{Total } SS - \text{Treatment } SS$$

$$= 42.94 - 40.72$$

$$= 2.22$$

Analysis of Variance of Dry Matter Production of Cassava from 5 Different Fertilizer Applications

Source of Variation	Degree of Freedom	Sum of Squares	Mean Square	F
Treatment	4	40.94	10.24	68.26**
Error	15	2.22	0.15	
Total	19	42.94		

** = Highly significant or significant at 1% level.

b. Since the computed F value is greater than the tabular F value of 4.89 at the 1% level,

we conclude that the difference between the treatments is highly significant. We also notice that because of the missing observations, the differences observed are highly significant as

noted in the high value of F.

c. The coefficient of variation is:

$$cv = \frac{\sqrt{\text{Error } MS}}{\text{Grand mean}} \times 100$$

$cv = 12.15$

d. We notice that with the missing observations, the efficiency of the experiment has gone up from 17.07% to 12.15%.

3. Seed yield of early maturing high protein soybean lines adapted to the Mid-Atlantic area of the U.S. were evaluated in field test in a randomized complete block with 3 replications. Each plot contained four 20-ft rows, 30 in. apart. Each plot was evaluated for seed yield, and the following data were recorded:

Seed Yield (bu/acre) of early maturing soybean lines

Line	Rep. I	Rep. II	Rep. III
CX797-115	32.2	33.5	33.3
CX797-21	35.8	36.2	36.8
CX804-3	34.5	35.4	36.6
K1085	37.2	36.4	38.3
K1091	39.8	36.2	40.2
Williams	37.8	38.2	41.1
Douglas	36.4	38.6	40.2

(a) Perform an analysis of variance
(b) Compute the coefficient of variation.
(c) Determine the relative efficiency of this experiment.

Solution:

a.

Line	Rep. I	Rep. II	Rep. III	Treatment Total	Treatment Mean
CX797-115	32.2	33.5	33.3	99.0	33.00
CX797-21	35.8	36.2	36.8	108.8	36.27
CX804-3	34.5	35.4	36.6	106.5	35.50
K1085	37.2	36.4	38.3	111.9	37.30
K1091	39.8	36.2	40.2	116.2	38.73
Williams	37.8	38.2	41.1	117.1	39.03
Douglas	36.4	38.6	40.2	115.2	38.40
Grand Total					774.70
Grand Mean					36.89

The correction factor is:

$$C = \frac{G^2}{rt} = \frac{(774.7)^2}{(3)(7)} = 28579$$

$$\text{Total } SS = \sum_{j=1}^{t} \sum_{i=1}^{r} X_{ij}^{2} - C$$

$$\text{Total } SS = [(32.2)^2 + (35.8)^2 + \ldots\ldots + (40.2)^2] - 28579$$

$$= 112.7$$

$$\text{Replication } SS = \frac{\sum R^2}{t} - C$$

$$= \frac{(253.7)^2 + (254.5)^2 + (266.5)^2}{7} - 28579$$

$$= 14.69$$

$$\text{Treatment } SS = \frac{\sum T_i^2}{r} - C$$

$$= \frac{(99)^2 + (108.8)^2 + \ldots + (115.2)^2}{3} - 28579$$

$$= 83.68$$

$$\text{Error } SS = \text{Total } SS - \text{Replication } SS - \text{Treatment } SS$$

$$= 112.7 - 14.69 - 83.68$$

$$= 14.31$$

Analysis of Variance of Seed Yield of Early Maturing Soybean Lines

Source of Variation	Degree of Freedom	Sum of Squares	Mean Square	F
Treatment	6	83.68	13.95	11.69**
Replication	2	14.69	7.35	
Error	12	14.31	1.19	
Total	20	112.68		

** = Highly significant or significant at 1% level.

Since the computed value of 11.69 is greater than the tabular value of 4.82 at the 1% level, we conclude that there is a significant difference between the yield of differing soybean seed lines.

b. The coefficient of variation is:

$$cv = \frac{\sqrt{\text{Error } MS}}{\text{Grand mean}} \times 100$$

$$cv = 2.84\%$$

c. The relative efficiency of the experiment is computed as:

$$R.E. = \frac{(r-1)s_B^2 + r(t-1)s_e^2}{(rt-1)s_e^2} \cdot \frac{[(r-1)(t-1)+1][t(r-1)+3]}{[(r-1)(t-1)+3][t(r-1)+1]}$$

$$R.E. = \frac{(3-1)(7.35) + 3(7-1)(1.19)}{(21-1)1.19} \cdot \frac{(13)(17)}{(15)(15)}$$

$$R.E. = 1.49$$

The results show that by using the randomized complete block design over the completely randomized design, the efficiency of the experiment has improved by 49%.

4. An ornamental horticulturist conducted a fertilizer experiment in a greenhouse where 5 fertilizer treatments (A, B, C, D, and E) were tested by arranging plants in a Latin-square design. Thus rows and columns in the table are rows and columns in the greenhouse. The data below shows the yield from the experiment.

A22	B23	C19	D12	E14
B20	C13	D16	E19	A18
C14	D10	E12	A26	B23
D19	E18	A20	B18	C14
E15	A24	B20	C17	D10

(a) Using a 1% level of significance, determine if the mean yields are not equal for the 5 fertilizers.
(b) Compute the coefficient of variation for the data.
(c) What can be said about the relative efficiency of the Latin-square design?

Solution:
a.

Row, Column, and Treatment Totals and Means for Fertilizer Experiment

Row	Column I	II	III	IV	V	Row Total	Treatment Total
1	A22	B23	C19	D12	E14	90	110A
2	B20	C13	D16	E19	A18	86	104B
3	C14	D10	E12	A26	B23	85	77C
4	D19	E18	A20	B18	C14	89	67D
4	E15	A24	B20	C17	D10	86	78E
Column Total	90	88	87	92	79		
Grand Total							436.00
Grand Mean							17.44

The correction factor is:

$$C = \frac{G^2}{t^2} = \frac{(436)^2}{25} = 7603.84$$

$$\text{Total } SS = \sum X_{ijk}^2 - C$$

Total $SS = \left[(22)^2 + (23)^2 + (19)^2 + \ldots (10)^2\right] - 7603.84$

$\qquad = 460.16$

Treatment $SS = \dfrac{\sum T^2}{k} - C$

$\qquad = 1/5 \left[(110)^2 + (104)^2 + (77)^2 + (67)^2 + (78)^2\right] - 7603.84$

$\qquad = 279.76$

Row $SS = \dfrac{\sum R_i^2}{r} - C$

$\qquad = 1/5 \left[(90)^2 + (86)^2 + (85)^2 + (89)^2 + (86)^2\right] - 7603.84$

$\qquad = 3.76$

Col. $SS = \dfrac{\sum C_j^2}{c} - C$

$\qquad = 1/5 \left[(90)^2 + (88)^2 + (87)^2 + (92)^2 + (79)^2\right] - 7603.84$

$\qquad = 19.76$

Error SS = Total SS - Treatment SS - Row SS - Col. SS

$\qquad = 460.16 - 279.76 - 3.76 - 19.76$

$\qquad = 156.88$

Analysis of Variance of Seed Yield of Early Maturing Soybean Lines

Source of Variation	Degree of Freedom	Sum of Squares	Mean Square	F
Treatment	4	279.76	69.94	5.35**
Row	4	3.76	0.94	
Column	4	19.76	4.94	
Error	12	156.88	13.07	
Total	24	460.16		

** = Highly significant or significant at 1% level.

a. The tabular value for the 1% level of significance is 5.99 and the computed value is 5.35. Thus, we conclude that at this level of significance there is no difference in the treatments.

b. The coefficient of variation is:

$$cv = \dfrac{\sqrt{\text{Error } MS}}{\text{Grand mean}} \times 100$$

$$cv = \frac{\sqrt{13.07}}{17.44} \times 100$$

$$cv = 20.72$$

c. The efficiency of the Latin square

$$R.E. = \frac{S_r^2 + S_c^2 + (t-1)S_e^2}{(t+1)S_e^2}$$

s_r^2 = row mean square

s_c^2 = column mean square

s_e^2 = error mean square

t = number of treatments.

$$R.E. = \frac{0.94 + 4.94 + (5-1)(13.07)}{(5+1)(13.07)}$$

$$R.E. = 0.72$$

Since the error degree of freedom is less than 20 we multiply by the k factor.

$$k = \frac{[(t-1)(t-2)+1][(t-1)^2+3]}{[(t-1)(t-2)+3][(t-1)^2+1]}$$

$$k = \frac{[(4)(3)+1][(4)^2+3]}{[(4)(3)+3][(4)^2+1]}$$

$$k = 0.97$$

$R.E. = (0.72)(0.97 = 0.69$. It appears from the results that the efficiency did not increase by using a Latin square design.

5. Plant breeders are interested in determining the spikelet initiation differences among 9 winter wheat cultivars. The number of spikelets per plant from a field experiment which followed a 3 x 3 balanced lattice design with 4 replications are given below. Each cultivar is given a treatment number and they are: Turkey (1), Pawnee (2), Scout (3), Larned (4), Newton (5), Hawk (6), Vona (7), HW 1010 (8), and Bounty 100 (9). The data collected from each replication are presented below.

Incomplete Block No.	Spikelet No. Rep. I			Incomplete Block No.	Spikelet No. Rep. II		
1	18.1(8)	18.4(6)	17.6(1)	4	18.2(8)	20.2(7)	16.5(9)
2	16.5(3)	18.7(5)	17.9(7)	5	15.2(3)	19.9(2)	17.8(1)
3	16.0(4)	18.0(2)	16.0(9)	6	17.8(6)	18.1(5)	16.4(4)
	Rep. III				Rep. IV		
7	17.1(8)	18.4(5)	18.6(2)	10	16.2(3)	15.9(4)	18.5(8)
8	16.2(4)	17.7(7)	16.9(1)	11	17.2(6)	18.9(2)	17.6(7)
9	16.5(3)	18.9(6)	16.2(9)	12	15.4(9)	18.9(5)	17.4(1)

(a) Perform the analysis of variance.
(b) From the analysis in (a), is it necessary to compute the adjusted treatment totals for all treatments?
(c) Compute the coefficient of variation.
(d) Compute the relative efficiency coefficient for this design. What do the results indicate?

Solution:

a.

Block No.				Block Total (B)	Treatment Number	Treatment Total (T)	Block Total (B_T)	W
		Rep. I						
1	18.1	18.4	17.6	54.1	1	69.7	209.5	0.9
2	16.5	18.7	17.9	53.1	2	75.4	210.7	13.2
3	16.0	18.0	16.0	50.0	3	64.4	208.2	-9.8
		Rep. Total		157.2	4	64.5	203.7	8.5
					5	74.1	211.2	7.3
		Rep. II			6	72.3	211.7	-0.1
4	18.2	20.2	16.5	54.9	7	73.4	212.5	0.0
5	15.2	19.9	17.8	52.9	8	71.9	213.7	-9.3
6	17.8	18.1	16.4	52.3	9	64.1	208.2	-11.0
		Rep. Total		160.1	Total (G)	629.8	1889.4	0.0
		Rep. III						
7	17.1	18.4	18.6	54.1				
8	16.2	17.7	16.9	50.8				
9	16.5	18.9	16.2	51.6				
		Rep. Total		156.5				
		Rep. IV						
10	16.2	15.9	18.5	50.6				
11	17.2	18.9	17.6	53.7				
12	15.4	18.9	17.4	51.7				
		Rep. Total		156.0				

$$W = kT - (k + 1)B_T + G$$

where

k = block size,

T, B_T, and G as defined above.

$$C = \frac{G^2}{(k)^2(k+1)}$$

$$C = \frac{(629.8)^2}{(9)(4)}$$

$$C = 11018$$

$$\text{Total } SS = \sum X_{ijk}^2 - C$$

$$\text{Total } SS = \left[(18.1)^2 + (18.4)^2 + (17.6)^2 + \ldots + (17.4)^2\right] - 11018$$

$$= 52.72$$

$$\text{Replication } SS = \frac{\sum R^2}{k^2} - C$$

$$= \frac{(157.2)^2 + (160.1)^2 + (156.5)^2 + (156)^2}{9} - 11018$$

$$= 1.12$$

$$\text{Treatment (unadj.) } SS = \frac{\sum T^2}{(k+1)} - C$$

$$= \frac{(69.7)^2 + (75.4)^2 + \ldots + (64.1)^2}{4} - 11018$$

$$= 40.73$$

$$\text{Block (adj.) } SS = \frac{\sum W^2}{(k^3)(k+1)}$$

$$= \frac{(0.9)^2 + (13.2)^2 + (-9.8)^2 + \ldots + (-11.0)^2}{(27)(4)}$$

$$= 5.53$$

$$\text{Intrablock error } SS = \text{Total } SS - \text{Replication } SS - \text{Treatment (unadj.)} SS - \text{Block (adj.)} SS$$

$$= 52.72 - 1.12 - 40.73 - 5.53$$

$$= 5.34$$

The mean square for each of the above sum of squares is:

$$\text{Treatment (unadj.) } MS = \frac{\text{Treatment (unadj.) } SS}{k^2 - 1}$$

Thus, the mean square for each source of variation is:

Treatment unadjusted mean square = 5.092
Block adjusted mean square = 0.692
Intrablock error mean square = 0.333

We now compute the adjusted treatment total ($\dot{T}$):

$$\dot{T} = T + \mu W$$

where μ equals:

$$\mu = \frac{\text{Block (adj.) } MS - \text{Intrablock error } MS}{k^2 [\text{Block (adj.) } MS]}$$

$\mu = 0.028$

$$\text{Treatment (adj.) } MS = \left[\frac{1}{(k+1)(k^2-1)}\right]\left[\sum \dot{T}^2 - \frac{G^2}{k^2}\right]$$

$$\text{Treatment (adj.) } MS = \left[\frac{1}{(4)(8)}\right]\left[44244 - 44072\right]$$

$$= 5.38$$

$$\text{Effective error } MS = (\text{Intrablock error } MS)(1 + k\,\mu)$$
$$= (0.333)(1 + (3)(0.028)$$
$$= 0.361$$

$$F = \frac{\text{Treatment (adj.) } MS}{\text{Effective error } MS}$$

$$F = \frac{5.77}{0.361} = 14.91$$

Analysis of Variance of Spikelet Initiation for 9 Winter Wheat Cultivars

Source of Variation	Degree of Freedom	Sum of Squares	Mean Square	F
Replication	3	1.12	0.37	
Block (Adj.)	8	5.53	0.69	
Treatment (adj.)	8	43.07	5.38	14.91**
Effective error	16	5.77	0.36	
Total	35	52.72		

** = Highly significant or significant at 1% level.

The tabular F (for 8 treatment degrees of freedom and 16 intrablock error degrees of freedom) is 2.59 and 3.89 for 5% and 1% levels of significance, respectively. The results suggests that there is a significant difference in the spikelet initiation among the 9 winter wheat cultivars.

b. Since the intrablock error mean square was less than the block (adj.) mean square, we therefore computed the adjusted treatment total ($\hat{T}$).

c. The coefficient of variation is:

$$cv = \frac{\sqrt{\text{Intrablock error } MS}}{\text{Grand Mean}} \times 100$$

$$= \frac{\sqrt{0.361}}{17.49} \times 100 = 3.43\%$$

d. The relative efficiency is computed as:

$$R.E. = \frac{\text{Block (adj.) } SS + \text{Intrablock error } SS}{(k-1)(k^2-1)(\text{Intrablock } MS)} \times 100$$

$$= \frac{5.53 + 5.776}{(2)(8)(0.361)} \times 100$$

$$= 195.74\%$$

The result indicates that the experimental precision has increased by 95.74% over a randomized complete block design.

6. An animal scientist is interested in evaluating the role of progestron in stimulating sexual receptivity in estrogen-treated gilts. She has used a 4 x 4 triple lattice design to conduct the experiment in which the 16 ovariectomized gilts were treated with estradil benzoate (EB). After EB treatment gilts were moved to an evaluation pen where bores were brought in. Gilts remained in the evaluation pen for 5 minutes, during which time the number of mounts attempted by the bore were recorded. The following data were collected from the experiment with 3 replications. The treatment numbers appear in parentheses.

Incomplete Block Number	Mounts, Number/5 min			
	Replication I			
1	7(01)	5(02)	4(03)	2(04)
2	5(05)	2(06)	1(07)	3(08)
3	4(09)	3(10)	2(11)	2(12)
4	1(13)	3(14)	1(15)	5(16)
	Replication II			
1	6(01)	6(05)	6(09)	2(13)
2	4(02)	2(06)	3(10)	3(14)
3	3(03)	3(07)	1(11)	2(15)
4	1(04)	1(08)	3(12)	5(16)
	Replication III			
1	7(01)	5(06)	4(11)	6(16)
2	4(05)	4(02)	1(15)	3(12)
3	5(09)	2(14)	4(03)	2(08)
4	1(13)	3(10)	2(07)	2(04)

(a) Perform the analysis of variance.

(b)   Estimate the gain in accuracy over randomized blocks.

Solution:

a.

Compute the treatment total (T) for all treatments. For example, the total for treatment number 1 is:

$$T_1 = 7 + 6 + 7 = 20$$

Calculate for each block, the C_b values where:

C_b = total (over all replications) of all treatments in the block - rB

where

r = number of replications.

B = block total.

the C_b value for block 1 in replication 1 is:

$$C_b = 20 + 13 + 11 + 5 - (3)(18) = -5$$

The C_b value for all the 15 blocks are computed and are shown in the following Table.

Block Number					Block Total	Treatment Number	Treatment Total	C_b
		Replication I						
1	7(01)	5(02)	4(03)	2(04)	18	1	20	-5
2	5(05)	2(06)	1(07)	3(08)	11	2	13	3
3	4(09)	3(10)	2(11)	2(12)	11	3	11	6
4	1(13)	3(14)	1(15)	5(16)	10	4	5	2
						5	15	6
						6	9	
		Replication II						
1	6(01)	6(05)	6(09)	2(13)	20	7	6	-6
2	4(02)	2(06)	3(10)	3(14)	12	8	6	3
3	3(03)	3(07)	1(11)	2(15)	9	9	15	1
4	1(04)	1(08)	3(12)	5(16)	10	10	9	5
						11	7	3
						12	8	
		Replication III						
1	7(01)	5(06)	4(11)	6(16)	22	13	4	-14
2	4(05)	4(02)	1(15)	3(12)	12	14	8	4
3	5(09)	2(14)	4(03)	2(08)	13	15	4	1
4	1(13)	3(10)	2(07)	2(04)	8	16	16	0
								-9

The replicate totals (R_c) of the C_b values are 6, 3, and -9. Note that these replicate totals sum to 0.

The correction factor is:

$$C = \frac{G^2}{(r)(k)^2}$$

$$= \frac{(156)^2}{(3)(4)^2}$$

$$= 507$$

$$\text{Total } SS = \sum X_{ij}^2 - C$$

$$\text{Total } SS = \left[(7)^2 + (5)^2 + (4)^2 + \ldots + (2)^2 \right] - 507$$

$$= 139$$

$$\text{Replication } SS = \frac{\sum R^2}{k^2} - C$$

$$= \frac{(50)^2 + (51)^2 + (55)^2}{16} - 507$$

$$= 0.875$$

$$\text{Treatment } SS \text{ (unadj.)} = \frac{\sum T^2}{r} - C$$

$$= \frac{(20)^2 + (13)^2 + \ldots + (16)^2}{3} - 507$$

$$= 114.33$$

$$\text{Block } SS \text{ (adj.)} = \frac{\sum C_b^2}{kr(r-1)} - \frac{\sum R_c^2}{k^2 r(r-1)}$$

$$= \frac{(-5)^2 + (2)^2 + \ldots + (0)^2}{(4)(3)(3-1)} - \frac{(6)^2 + (3)^2 + (-9)^2}{(4)^2(3)(3-1)}$$

$$= 13.60$$

Intrablock error SS = TSS - Rep. SS - Treatment (unadj.) SS - Block (adj.) SS

$$= 139.0 - 0.875 - 114.33 - 13.60$$

$$= 10.19$$

$$\text{Block } MS \text{ (adj.)} = \frac{\text{Block } SS \text{ (adj.)}}{r(k-1)}$$

$$= \frac{13.60}{9}$$

$$= 1.51$$

$$\text{Intrablock Error } MS = \frac{\text{Intrablock Error } SS}{(k-1)(rk-k-1)}$$

$$= \frac{10.19}{(4-1)\left[(3)(4)-4-1\right]}$$

$$= 0.485$$

If the block (adj.) mean square is less than the intrablock error mean square, μ is assumed to be 0 and there is no need for adjustments for the block effects. However, for the present example the block (adj.) mean square is greater than the intrablock error mean square, so the adjustment factor is calculated as:

$$\mu = \frac{\text{Block (adj.) } MS - \text{Intrablock error } MS}{k\,(r-1)\left[\text{Block (adj.) } MS\right]}$$

$$\mu = \frac{1.51 - 0.485}{4(3-1)(1.51)}$$

$$= 0.085$$

Calculate the adjusted treatment SS as:

$$\text{Treatment (adj.) } SS = \text{Treatment (unadj.) } SS - k\,(r-1)\,\mu\left\{\left[\frac{r}{(r-1)(1+k\mu)}\right]B_u - B_a\right\}$$

where

B_u = unadjusted sum of squares for blocks within replications

B_a = adjusted sum of squares for blocks within replications.

we first have to calculate B_u which is defined as:

$$B_u = \frac{\sum B^2}{k} - \frac{\sum R^2}{k^2}$$

where

B = block total.

R = replication total.

k = number of blocks.

$$B_u = \frac{(18)^2 + (11)^2 (11)^2 + \ldots + (8)^2}{4} - \frac{(50)^2 + (51)^2 + (55)^2}{(4)^2}$$

$$= 55.125$$

Substituting the value of B_u and B_a into the adjusted treatment SS, we will have:

$$\text{Treatment (adj.) } SS = 114.33 - 4(2)(0.085)\left\{\left[\frac{3}{(2)(1 +(4)(0.085)}\right](55.125) - (13.60)\right\}$$

$$= 81.64$$

$$\text{Treatment } MS \text{ (adj.)} = \frac{\text{Treatment } SS \text{ (adj.)}}{k^2 - 1}$$

$$= \frac{81.64}{15}$$

$$= 5.44$$

Analysis of Variance of an Experiment on the Role of Progestron in Stimulating Sexual Receptivity in Estrogen Treated Gilts

Source of Variation	Degrees of Freedom	Sum of Squares	Mean Square	F
Replication	2	0.87		
Block (adj.)	9	13.60	1.51	
Treatment (unadj.)	(15)	114.33		
Treatment (adj.)	15	81.64	5.44	11.33**
Intrablock error	21	10.19	0.48	
Total	47	139.0		

**= highly significant

b.

Estimate the gain in accuracy over randomized blocks. This is accomplished by comparing the error mean square with the effective error variance. The error mean square in the randomized block is the pooled mean square for blocks and intrablock error as shown below:

$$\text{Randomized Block } MS = \frac{\text{Block } SS \text{ (adj.)} + \text{Intrablock error } SS}{(k^2 - 1)(r - 1)}$$

$$\text{Randomized block } MS = \frac{13.60 + 10.19}{30}$$

$$= 0.79$$

The effective error variance is computed as:

$$\text{Intrablock error } MS \left[1 + \frac{rk\mu}{(k + 1)}\right]$$

$$\text{Effective error variance} = 0.48\left[1 + \frac{(3)(4)(0.085)}{(4 + 1)}\right]$$

$$= 0.577$$

Thus, the relative accuracy is:

$$\frac{0.79}{0.577} = 136.9\%$$

7. Suppose the animal scientist in the previous exercise chose to conduct her experiment as a case with repetition. As you recall from your reading in this chapter, when the replicates contained in the basic plan are repeated p times so that the total number of replication is $r = np$, then we have a case of data analysis with repetition. For the present case, the basic plan of the first 2 replicates have been repeated. This changes a triple lattice to a quadruple lattice. The data collected from this experiment are shown below.

Incomplete Block No.	Mounts, Number/5 min				Incomplete Block No.	Mounts, Number/5 min			
	Rep. I					Rep. II			
1	7(01)	5(02)	4(03)	2(04)	5	6(01)	6(05)	5(09)	2(13)
2	5(05)	2(06)	1(07)	3(08)	6	4(02)	2(06)	3(10)	3(14)
3	4(09)	3(10)	2(11)	2(12)	7	3(03)	3(07)	1(11)	2(15)
4	1(13)	3(14)	1(15)	5(16)	8	1(04)	1(08)	3(12)	5(16)
	Rep. III					Rep. IV			
1	6(01)	4(02)	3(03)	1(04)	5	4(01)	5(05)	5(09)	1(13)
2	4(05)	1(06)	2(07)	4(08)	6	3(02)	3(06)	4(10)	4(14)
3	5(09)	2(10)	1(11)	2(12)	7	2(03)	4(07)	2(11)	3(15)
4	3(13)	5(14)	2(15)	4(16)	8	1(04)	3(08)	2(12)	4(16)

(a)   Perform the analysis of variance.
(b)   Estimate the gain in accuracy of this design over randomized blocks.

Solutions:

a.

Compute the replication totals, the block totals (B), and the grand total (G) as shown in Table below.

Compute the treatment total (T) for all treatments. For this example, the total for treatment number one is:

$$T_1 = 7 + 6 + 6 + 4 = 23$$

Block Number					Block Total	Treatment Number	Treatment Total
	Replication I						
1	7(01)	5(02)	4(03)	2(04)	18	1	23
2	5(05)	2(06)	1(07)	3(08)	11	2	16
3	4(09)	3(10)	2(11)	2(12)	11	3	12
4	1(13)	3(14)	1(15)	5(16)	<u>10</u>	4	5
					50	5	20
	Replication II					6	8
5	6(01)	6(05)	6(09)	2(13)	14	7	10
6	4(02)	2(06)	3(10)	3(14)	11	8	11

7	3(03)	3(07)	1(11)	2(15)	10	9	19
8	1(04)	1(08)	3(12)	5(16)	<u>14</u>	10	12
					49	11	6
		Replication III				12	9
1	6(01)	4(02)	3(03)	1(04)	19	13	7
2	4(05)	1(06)	2(07)	4(08)	12	14	15
3	5(09)	2(10)	1(11)	2(12)	9	15	8
4	3(13)	5(14)	2(15)	4(16)	<u>10</u>	16	18
					50		
		Replication III					
5	4(01)	5(05)	5(09)	1(13)	15		
6	3(02)	3(06)	4(10)	4(14)	14		
7	2(03)	4(07)	2(11)	3(15)	11		
8	1(04)	3(08)	2(12)	4(16)	<u>10</u>		
					50		
Grand Total (G)					**2923**		

Determine the degrees of freedom associated with each source of variation.

Replication *d.f.*	$= r - 1 = 3$
Treatment (unadj.) *d.f.*	$= k^2 - 1 = 15$
Blocks within replications (adj.) *d.f.*	$= r(k - 1) = 12$
Component (a) *d.f.*	$= n(p - 1)(k - 1) = 6$
Component (b) *d.f.*	$= n(k - 1) = 6$
Intrablock error *d.f.*	$= (k - 1)(rk - k - 1) = 33$
Total *d.f.*	$= (rk^2 - 1) = 63$

where

r = total number of replications

k = block size

n = number of replication in the base design

p = number of repetition.

The correction factor is:

$$C = \frac{G^2}{(n)(p)(k^2)}$$

$$= \frac{(199)^2}{(2)(2)(16)} = 618.77$$

$$\text{Total } SS = \sum X^2 - C$$

$$= \left[(7)^2 + (5)^2 + (4)^2 + \ldots (4)^2 \right] - 618.77$$

$$= 148.23$$

$$\text{Rep. } SS = \frac{\sum R^2}{k^2} - C$$

$$= \frac{(50)^2 + (49)^2 + (50)^2 + (50)^2}{16} - 618.77$$

$$= 0.05$$

$$\text{Treatment } SS \text{ (unadj.)} = \frac{\sum T^2}{(n)(p)} - C$$

$$= \frac{2923}{(2)(2)} - 618.77$$

$$= 111.98$$

For each block in each repetition, calculate the B_T value, which is the sum of block totals over all replications in that repetition. For example, the B_T value for block 1 from replications I and III is:

$$B_T = 18 + 14 = 32$$

Calculate for each block, the C_b values as:

$$C_b = \sum T - nB_T$$

where

T = Treatment total
n = number of replications in the base design
B_T = sum of block totals over all replications.

For example, the C_b value for block 1 in replication 1 which contains treatments 1, 2, 3, 4, and 5 is:

$$C_b = 23 + 16 + 12 + 5 - (2)(32) = -8$$

The C_b value for all the blocks are computed and are shown in Table

B_T and C_b Values for Pairs of Blocks Containing the Same Set of Treatments

Block Number	Block Totals		Sum (B_T)	C_b
	Repetition 1			
	Rep. I	Rep. III		
1	18	14	32	-8
2	11	11	22	5
3	11	10	21	4
4	10	14	24	0
Total	50	49	99	1

	Repetition 2			
	Rep. II	Rep. IV		
1	19	15	34	1
2	12	14	26	-1
3	9	11	20	-4
4	10	10	20	3
Total	50	50	100	-1

Compute the block SS which contains 2 components, (a) and (b).

Component (a) $SS =$ Total - Rows - Columns

Where:

$$\text{Total} = \frac{\sum B^2}{k} - \frac{\sum R^2}{pk^2}$$

$$\text{Rows} = \frac{\sum B_T^2}{pk} - \frac{\sum R^2}{pk^2}$$

$$\text{Columns} = \frac{\sum N^2}{k^2} - \frac{\sum R^2}{pk^2}$$

The term B is the block total, R is the sum of B_T values for each repetition, N is the sum of block totals for each replications, and k and p are as defined earlier. For our example, the values are:

$$\text{Total} = \frac{(18)^2 + (11)^2 + \ldots + (11)^2 + (10)^2}{4} - \frac{(99)^2 + (100)^2}{32}$$

$$= 32.97$$

$$\text{Rows} = \frac{(32)^2 + (22)^2 + \ldots + (20)^2 + (20)^2}{8} - \frac{(99)^2 + (100)^2}{32}$$

$$= 25.84$$

$$\text{Columns} = \frac{(50)^2 + (49)^2 + (50)^2 + (50)^2}{16} - \frac{(99)^2 + (100)^2}{32}$$

$$= 0.03$$

From the above, we compute the component (a) as follows:

Component (a) $SS = 32.97 - 25.84 - 0.03$

$$= 7.10$$

Component (b) $SS = \dfrac{\sum C_b^2}{kr\,(n-1)} - \dfrac{\sum R_c^2}{k^2 r\,(n-1)}$

where

R_c = the total of C_b values over all blocks in a repetition.

$$\text{Component } (b)\ SS = \dfrac{(-8)^2+(5)^2+\dots+(-4)^2+(3)^2}{16} - \dfrac{(1)^2+(-1)^2}{64}$$

$$= 8.22$$

Calculate the adjusted sum of squares for blocks within replications as:

Block SS (adj.) = Component $(a)SS$ + Component $(b)SS$

$$= 7.10 + 8.22$$

$$= 15.32$$

Calculate the intrablock error sum of squares as:

Intrablock error SS = Total SS - Rep. SS - Treatment SS (unadj) - Block SS (adj.)

$$= 148.23 - 0.05 - 111.98 - 15.32$$

$$= 20.88$$

Calculate the mean square for block (adj.) and the intrablock error as:

Block MS (adj.) $= \dfrac{\text{Block } SS \text{ (adj.)}}{r\,(k-1)}$

$$= \dfrac{15.32}{12} = 1.28$$

Intrablock Error $MS = \dfrac{\text{Intrablock Error } SS}{(k-1)(rk-k-1)}$

$$= \dfrac{20.88}{(4-1)\left[(4)(4)-4-1\right]} = 0.63$$

Calculate μ, the weighting factor, to adjust the treatment totals. If the block (adj.) mean square is the pooled mean square for the blocks, then the weighting factor is:

$$\mu = \dfrac{p\left[\text{Block (adj.) } MS - \text{Intrablock error } MS\right]}{k\left[(r-p)\text{Block (adj.) } MS + (p-1)\text{Intrablock error } MS\right]}$$

$$= 0.10$$

Keep in mind that If block (adj.) mean square is less than the intrablock error mean square, μ is assumed to be 0 and there is no need for adjustments for the block effects. Since the opposite holds true for the present case, each treatment total is adjusted by applying a correction for each group of blocks in which the treatment appears. Given that a correction is necessary, calculate the adjusted treatment sum of squares as:

$$\text{Treatment } SS \text{ (adj.)} = \text{Treatment } SS \text{ (unadj.)} - k(n-1)\mu\left[\frac{n}{(n-1)(1+k\mu)}B_u - B_a\right]$$

where

$\quad B_u$ = unadjusted sum of squares for component b of the blocks
$\quad B_a$ = adjusted sum of squares for component b of the block.

We have already computed $B_u \cdot$ which is 25.84 and was calculated as the rows sum of squares when computing component (a). B_a is the component (b) sum of squares and is 8.22. Thus, the adjusted treatment sum of squares is:

$$\text{Treatment } SS \text{ (adj)} = 111.98 - 4(1)(0.10)\left\{\left[\frac{2}{(1)(1+4)(0.10)}\right]25.84 - 8.22\right\}$$

$$= 73.92$$

Calculate the adjusted treatment mean square as:

$$\text{Treatment } MS \text{ (adj.)} = \frac{\text{Treatment } SS \text{ (adj.)}}{(k^2 - 1)}$$

$$= \frac{73.92}{15} = 4.93$$

Compute the F value as:

$$F = \frac{\text{Treatment (adj.) } MS}{\text{Effective error } MS}$$

$$= \frac{4.93}{0.63} = 7.83$$

Analysis of Variance of a Case with Repetition of an Experiment on the Role of Progestron in Stimulating Sexual Receptivity in Estrogen Treated Gilts

Source of Variation	Degrees of Freedom	Sum of Squares	Mean Square	F
Replication	3	0.05	0.016	
Block (adj.)	12	15.32	1.28	
Component (a)	(6)	7.10		
Component (b)	(6)	8.22		
Treatment (unadj.)	(15)	111.98		
Treatment (adj.)	15	73.92	4.93	7.81**
Intrablock error	33	20.88	0.63	
Total	63	148.23		

**= highly significant

The results indicate that there is a highly significant difference among the treatments.

b.

Compute the error variance or the effective error mean square for the difference between 2 treatment means as shown below.

1. The error mean square for 2 treatments in the same block:

$$\text{Error } MS = MS \text{ error } [1 + (n - 1)\mu]$$
$$= 0.63[1+(2-1)0.10]$$
$$= 0.69$$

2. The error mean square for 2 treatments not the in the same block:

$$\text{Error } MS = MS \text{ error } (1 + n \mu)$$
$$= 0.63[1 + 2(0.10)$$
$$= .76$$

3. The average effective error mean square is computed as:

$$\text{Average Error } MS = MS \text{ error} \left[1 + \frac{nk\mu}{k + 1}\right]$$
$$= .63\left[1 + \frac{2\,(4)(0.10)}{4 + 1}\right]$$
$$= 0.73$$

Calculate the relative efficiency ($R.E.$) to estimate the precision relative to randomized complete block designs as:

$$R.E. = \left[\frac{\text{Block (adj.) } SS + \text{Intrablock error } SS}{r\,(k - 1) + (k - 1)(rk - k - 1)}\right]\left[\frac{100}{\text{Error } MS}\right]$$

1.
$$R.E. = \left[\frac{15.32 + 20.88}{4(3) + (3)(16 - 4 - 1)}\right]\left[\frac{100}{0.69}\right]$$
$$= 116.59\%$$

2.
$$R.E. = \left[\frac{15.32 + 20.88}{4(3) + (3)(16 - 4 - 1)}\right]\left[\frac{100}{0.76}\right]$$
$$= 105.85\%$$

3.
$$R.E. = \left[\frac{15.32 + 20.88}{4(3) + (3)(16 - 4 - 1)}\right]\left[\frac{100}{0.73}\right]$$
$$= 110.19\%$$

It appears that the relative efficiency of this experiment is higher than an experiment conducted as a randomized design.

Chapter 5

TWO-FACTOR EXPERIMENTAL DESIGNS

1. An animal scientist wishes to conduct a split-plot design experiment where 2 factors (2 breeds of swine and 4 different diets) are to be arranged in 4 blocks. The objective of the experiment is to see if the 4 diets are different from one another. Show the layout of this experiment, keeping the objective in mind.

Solution:

Diet	Breed 1 (B_1)	Breed 2 (B_2)
D_1	B_1D_1	B_2D_1
D_2	B_1D_2	B_2D_2
D_3	B_1D_3	B_2D_3
D_4	B_1D_4	B_2D_4

2. The omnivorous looper, larva of the moth *Sabulodes aegrotata* (Guenee) is a sporadic pest of California's avocado. Researchers are testing 4 chemicals to control the omnivorous looper in avocado orchards. The treatments were assigned to the subplots, whereas the concentration rates of active ingredients were assigned to the main plots. The results of the experiment are shown below.

Insecticide used for omnivorous looper control in San Diego County

Treatment	No. of larvae per tree sample at 14-day posttreatment interval		
	Rep. I	Rep. II	Rep. III
Active Ingredient 1 lb/ac			
Dylox 80SP	8	10	11
Kryocide 8F	12	9	10
Lannate L	2	4	3
Orthene 75SP	1	2	4
Control	14	18	20
Active Ingredient 2 lb/ac			
Dylox 80SP	5	9	8
Kryocide 8F	6	4	5
Lannate L	1	3	4
Orthene 75SP	1	2	4
Control	17	19	24
Active Ingredient 4 lb/ac			
Dylox 80SP	3	3	5
Kryocide 8F	5	4	7
Lannate L	2	1	3
Orthene 75SP	1	2	1
Control	12	15	22

(a) Perform the analysis of variance on the data.
(b) What conclusions can be drawn from this analysis.
(c) Determine the degree of precision with which the treatments are compared.

Solution:

a.

The Replication x Active Ingredient (Factor A) Totals Computed from Data

Replication	Active Ingredient Rates			Rep. Total (R)
	1 lb/ac	2 lb/ac	4 lb/ac	
I.	37	30	23	90
II.	43	37	25	105
III.	48	45	38	131
Active Ingredient Total (A)	128	112	86	
Grand Total (G)				326

The Insecticide x Active Ingredient (Factor B) Totals

Treatment	No. of Larvae Total (AB)			Treatment Total (B)
	Active Ingredient rate, lb/acre			
	1 lb	2 lb	4 lb	
Dylox 80SP	16	22	24	62
Kryocide *F	23	17	22	62
Lannate L	5	8	10	23
Orthene 75SP	3	6	9	18
Control	43	52	66	161

Determine the degrees of freedom associated with each source of variation. The following sources of variation and the degrees of freedom are identified:

$$\text{Replication } d.f. = r - 1 = 2$$

$$\text{Main plot factor } (A) d.f. = a - 1 = 2$$

$$\text{Error } (a) \ d.f. = (r - 1)(a - 1) = 4$$

$$\text{Subplot factor } (B) \ d.f. = b - 1 = 4$$

$$A \times B \ d.f. = (a - 1)(b - 1) = 8$$

$$\text{Error } (b) \ d.f. = a(r - 1)(b - 1) = 24$$

$$\text{Total } d.f. = rab - 1 = 44$$

The correction factor is:

$$C = \frac{G^2}{rab}$$

$$= \frac{(326)^2}{3 \times 3 \times 5} = 2361.69$$

Total $SS = \sum X^2 - C$

$$= [(8)^2 + (12)^2 + \ldots + (22)^2] - 2361.69$$

$$= 1744.31$$

Replication $SS = \frac{\sum R^2}{ab} - C$

$$= \frac{(90)^2 + (105)^2 + (131)^2}{15} - 2361.69$$
$$= 57.38$$

SS_A (Nitrogen) $= \dfrac{\sum A^2}{rb} - C$

$$= \frac{(128)^2 + (112)^2 + (86)^2}{3 \times 5} - 2361.69$$

$$= 59.91$$

Error $SS(a) = \dfrac{\sum (RA)^2}{b} - (C + \text{Rep. } SS + SS_A)$

$$= \frac{(37)^2 + (30)^2 + \ldots + (38)^2}{5} - (2361.69 + 57.38 + 59.91)$$

$$= 3.82$$

SS_B (Hybrid) $= \dfrac{\sum B^2}{ra} - C$

$$= \frac{(62)^2 + (62)^2 + (23)^2 + (18)^2 + (161)^2}{3 \times 3} - 2361.69$$

$$= 1467.42$$

$$SS_{AB} \text{ (Nitrogen x Hybrid)} = \frac{\sum (AB)^2}{r} - (C + SS_A + SS_B)$$

$$= \frac{(16)^2 + (22)^2 + \ldots + (66)^2}{3} - (2361.69 + 59.91 + 1467.42)$$

$$= 58.31$$

Error $SS\,(b) = TSS$ - All other sum of squares

$$= 1744.31 - (57.38 + 59.91 + 3.82 + 1467.42 + 58.31)$$

$$= 97.47$$

Finally, the ANOVA Table presents the completed results

Analysis of Variance of Insecticide Use in a Split-Plot Design

Source of Variation	Degree of Freedom	Sum of Squares	Mean Square	F
Replication	2	57.38		
Active ingredient (A)	2	59.91	29.95	31.53**
Error (a)	4	3.82	0.95	
Insecticide (B)	4	1467.42	366.86	93.11**
Active ingredient x Insecticide ($A \times B$)	8	58.31	7.29	1.85ns
Error (b)	24	94.47	3.94	
Total	44	1744.31		

ns = not significant, ** = significant at 1% level.

b. The analysis shows that the number of larvae found per sample tree were significantly affected by both the concentration of active and ingredients and the type of insecticide used.

c. The coefficient of variation is computed for the main as well as the subplot as:

$$cv = \frac{\sqrt{\text{Error }(a)\ MS}}{\text{Grand Mean}} \times 100$$

$$= \frac{\sqrt{0.95}}{7.24} \times 100 = 13.39\%$$

$$cv = \frac{\sqrt{\text{Error }(b)\ MS}}{\text{Grand Mean}} \times 100$$

$$= \frac{\sqrt{3.94}}{7.24} \times 100 = 27.41\%$$

As can be seen the subplot precision is greater than the mainplot. This is a feature of the split-plot design.

3. In a study to determine the viability of triticale as a feedstuff for poultry, animal researchers conducted a split-plot experiment. They gathered the following data on the weight gain of the birds fed triticale and control.

Cumulative growth of broilers fed control or triticale diets

Diet	Weight gain over days(lb/bird)			
	Rep. I	Rep. II	Rep. III	Rep. IV
7 Days				
Control	0.25	0.31	0.22	0.28
Low-triticale	0.26	0.28	0.29	0.27
Medium-triticale	0.22	0.21	0.24	0.22
14 Days				
Control	0.65	0.70	0.59	0.68
Low-triticale	0.73	0.75	0.79	0.69
Medium-triticale	0.70	0.69	0.65	0.68
21 Days				
Control	1.25	1.31	1.24	1.27
Low-triticale	1.29	1.28	1.29	1.30
Medium-triticale	1.22	1.21	1.23	1.22
28 Days				
Control	1.35	1.33	1.29	1.28
Low-triticale	1.46	1.38	1.40	1.39
Medium-triticale	1.42	1.31	1.34	1.36

(a) Perform the analysis of variance on the data.
(b) Compute the 2 coefficients of variation, one corresponding to the main plot and another to the subplot analysis.

Solution:

a.

The Replication x Duration (Factor A) Totals Computed from Data

Replication	Duration				Rep. Total (R)
	7 Days	14 Days	21 Days	28 Days	
I.	0.73	2.08	3.76	4.23	10.80
II.	0.80	2.14	3.80	4.02	10.76
III.	0.75	2.03	3.76	4.03	10.57
IV	0.77	2.05	3.79	4.03	10.64
Duration Total (A)	3.05	8.30	15.11	16.31	
Grand Total (G)					42.77

The Diet x Duration (Factor B) Totals

Diet	Duration				Diet Total (B)
	7 Days	14 Days	21 Days	28 Days	
Control	1.06	2.62	5.07	5.25	14.00
Low-triticale	1.10	2.96	5.16	5.63	14.85
Medium-triticale	0.89	2.72	4.88	5.43	13.92

Determine the degrees of freedom associated with each source of variation. The following sources
of variation and the degrees of freedom are identified:

$$
\begin{array}{llll}
\text{Replication } d.f. & = r - 1 & = 3 \\
\text{Main plot factor } (A)d.f. & = a - 1 & = 3 \\
\text{Error } (a) \ d.f. & = (r - 1)(a - 1) & = 9 \\
\text{Subplot factor } (B) \ d.f. & = b - 1 & = 2 \\
A \ \text{x} \ B \ d.f. & = (a - 1)(b - 1) & = 6 \\
\text{Error } (b) \ d.f. & = a \ (r - 1)(b - 1) & = 24 \\
\text{Total } d.f. & = r \ ab - 1 & = 47 \\
\end{array}
$$

The correction factor is:

$$
C = \frac{G^2}{rab}
$$

$$
= \frac{(42.77)^2}{4 \times 4 \times 3} = 38.11
$$

$$
\text{Total } SS = \sum X^2 - C
$$

$$
= [(0.25)^2 + (0.31)^2 + \ldots + (1.36)^2] - 38.11
$$

$$
= 9.69
$$

$$
\text{Replication } SS = \frac{\sum R^2}{ab} - C
$$

$$
= \frac{(10.80)^2 + (10.76)^2 + (10.57)^2 + (10.64)^2}{12} - 38.11
$$

$$
= 0.003
$$

$$
SS_A \text{ (Duration)} = \frac{\sum A^2}{rb} - C
$$

$$
= \frac{(3.05)^2 + (8.30)^2 + (15.11)^2 + (16.31)^2}{12} - 38.11
$$

$$
= 9.60
$$

$$\text{Error } SS\,(a) = \frac{\sum (RA)^2}{b} - (C + \text{Rep. } SS + SS_A)$$

$$= \frac{(0.73)^2 + (2.08)^2 + \ldots + (4.03)^2}{3} - (38.11 + 0.003 + 9.60)$$

$$= 0.01$$

$$SS_B \text{ (Diet)} = \frac{\sum B^2}{ra} - C$$

$$= \frac{(14.0)^2 + (14.85)^2 + (13.92)^2}{4 \times 4} - 38.11$$

$$= 0.03$$

$$SS_{AB} \text{ (Duration x Diet)} = \frac{\sum (AB)^2}{r} - (C + SS_A + SS_B)$$

$$= \frac{(1.06)^2 + (2.62)^2 + \ldots + (5.43)^2}{4} - (38.11 + 9.60 + 0.03)$$

$$= 0.02$$

$$\text{Error } SS\,(b) = TSS \; - \text{ All other sum of squares}$$

$$= 9.69 - (0.003 + 9.60 + 0.01 + 0.03 + 0.02)$$

$$= 0.03$$

Finally, the ANOVA Table presents the completed results

Analysis of Variance of Growth of Broilers Fed Triticale in a Split-Plot Design

Source of Variation	Degree of Freedom	Sum of Squares	Mean Square	F
Replication	3	0.003	0.0010	
Duration (A)	3	9.60	3.2000	2618.2**
Error (a)	9	0.01	0.0012	
Diet (B)	2	0.03	0.0165	18.0**
Duration x Diet (A x B)	6	0.02	0.0028	3.1*
Error (b)	24	0.03	0.0009	
Total	47	9.69		

* = significant at the 5% level, ** = significant at 1% level.

b. The results show that the weight gain was significantly affected by the diets. Furthermore, the weight gain was affected by the feeding duration.

c. The coefficient of variation is computed for the main as well as the subplot as:

$$cv = \frac{\sqrt{\text{Error } (a) \text{ } MS}}{\text{Grand Mean}} \times 100$$

$$= \frac{\sqrt{0.0012}}{0.89} \times 100 = 3.89\%$$

$$cv = \frac{\sqrt{\text{Error } (b) \text{ } MS}}{\text{Grand Mean}} \times 100$$

$$= \frac{\sqrt{0.0009}}{0.89} \times 100 = 3.37\%$$

As can be seen the subplot precision is greater than the mainplot. This is a feature of the split-plot design.

4. The western flower thrip, *Frankliniella occidentalis* (Pergande), is a major threat to the floricultural crops around the world. Researchers have conducted a split-plot experiment in which biological control has been used to control the thrips. In this experiment the objective is to determine if using predaceous mites such as *A. cucumeris* and *A.barkeri*, separately and in combination, would reduce the thrip population. Using chrysanthemum plants, the researchers have gathered the following data.

Impact of predaceous mites on the number of western flower thrips infesting chrysanthemum

Replication	Day	Amblyseius cucumeris	Amblyseius barkeri	A. cucumeris + A.barkeri	Control
I	Day 10	3	2	1	7
	Day 17	2	4	2	9
	Day 24	1	2	2	21
	Day 31	6	3	8	38
II	Day 10	2	1	2	9
	Day 17	3	4	1	10
	Day 24	5	2	3	25
	Day 31	3	5	7	34
III	Day 10	1	3	1	8
	Day 17	3	5	1	14
	Day 24	2	1	3	23
	Day 31	5	2	9	32
IV	Day 10	2	4	2	10
	Day 17	1	2	2	11
	Day 24	4	5	5	21
	Day 31	6	6	10	39

(a) Analyze the data.
(b) Compute the coefficient of variation for the 2 categories of the data.

Solution:

a.

<u>The Replication x Biological Control (Factor *A*) Totals Computed from Data</u>

| | Biological Controls | | | | |
Replication	*Amblyseius cucumeris*	*Amblyseius barkeri*	*A. cucumeris + A. barkeri*	Control	Rep. Total (*R*)
I.	12	11	13	75	111
II.	13	12	13	78	116
III.	11	11	14	77	113
IV	13	17	19	81	130
Duration Total (*A*)	49	51	59	311	
Grand Total (*G*)					470

<u>Duration x Biological Control (Factor *B*) Table of Totals</u>

| | Biological Controls | | | | |
Duration	*Amblyseius cucumeris*	*Amblyseius barkeri*	*A. cucumeris + A. barkeri*	Control	Duration Total (*B*)
Day 10	8	10	6	34	58
Day 17	9	15	6	44	74
Day 24	12	10	13	90	125
Day 31	20	16	34	143	213

Determine the degrees of freedom associated with each source of variation. The following sources

of variation and the degrees of freedom are identified:

Replication *d.f.*	$= r - 1$	$= 3$
Main plot factor (*A*) *d.f.*	$= a - 1$	$= 3$
Error (*a*) *d.f.*	$= (r - 1)(a - 1)$	$= 9$
Subplot factor (*B*) *d.f.*	$= b - 1$	$= 3$
A x *B* *d.f.*	$= (a - 1)(b - 1)$	$= 9$
Error (*b*) *d.f.*	$= a(r - 1)(b - 1)$	$= 36$
Total *d.f.*	$= r\,ab - 1$	$= 63$

The correction factor is:

$$C = \frac{G^2}{rab}$$

$$= \frac{(470)^2}{4 \times 4 \times 4} = 3451.56$$

Total $SS = \sum X^2 - C$

$$= [(3)^2 + (2)^2 + \ldots + (39)^2] - 3451.56$$

$$= 5274.44$$

$$\text{Replication } SS = \frac{\sum R^2}{ab} - C$$

$$= \frac{(111)^2 + (116)^2 + (113)^2 + (130)^2}{16} - 3451.56$$

$$= 13.81$$

$$SS_A \text{ (Biological Controls)} = \frac{\sum A^2}{rb} - C$$

$$= \frac{(49)^2 + (51)^2 + (59)^2 + (311)^2}{16} - 3451.56$$

$$= 3123.69$$

$$\text{Error } SS\,(a) = \frac{\sum (RA)^2}{b} - (C + \text{Rep. } SS + SS_A)$$

$$= \frac{(12)^2 + (11)^2 + \ldots + (81)^2}{4} - (3451.56 + 13.81 + 3123.69)$$

$$= 3.94$$

$$SS_B \text{ (Duration)} = \frac{\sum B^2}{ra} - C$$

$$= \frac{(58)^2 + (74)^2 + (125)^2 + (213)^2}{4 \times 4} - 3451.56$$

$$= 913.06$$

$$SS_{AB} \text{ (Bio logical Controls x Duration)} = \frac{\sum (AB)^2}{r} - (C + SS_A + SS_B)$$

$$= \frac{(8)^2 + (10)^2 + \ldots + (143)^2}{4} - (3451.56 + 3123.69 + 913.06)$$

$$= 1113.69$$

Error $SS(b) = TSS$ - All other sum of squares

$$= 5274.44 - (13.81 + 3123.69 + 3.94 + 913.06 + 1113.69)$$

$$= 106.25$$

Finally, the ANOVA Table presents the completed results

Analysis of Variance of Growth of Broilers Fed Triticale in a Split-Plot Design

Source of Variation	Degree of Freedom	Sum of Squares	Mean Square	F
Replication	3	13.81	4.60	
Biological controls (A)	3	3123.69	1041.23	2366.4**
Error (a)	9	3.94	0.44	
Duration (B)	3	913.06	304.35	103.1**
Biological controls x Duration (A x B)	9	1113.69	123.74	41.9**
Error (b)	36	106.25	2.95	
Total	63	5274.44		

** = significant at 1% level.

b. The results show that the biological controls as well as the combination of biological controls and the duration of the use of biological controls have significantly affected the reduction of thrips.

c. The coefficient of variation is computed for the main as well as the subplot as:

$$cv = \frac{\sqrt{\text{Error }(a)\text{ MS}}}{\text{Grand Mean}} \times 100$$

$$= \frac{\sqrt{0.44}}{7.34} \times 100 = 8.99\%$$

$$cv = \frac{\sqrt{\text{Error }(b)\text{ MS}}}{\text{Grand Mean}} \times 100$$

$$= \frac{\sqrt{2.95}}{7.34} \times 100 = 23.43\%$$

As can be seen the subplot precision is greater than the mainplot. This is a feature of the split-plot design.

5. Suppose in the above experiment, the researchers lost some of the observations as shown in the table below. The missing data is shown as (-) in each column of the data set.

Impact of predaceous mites on the number of western flower thrips infesting chrysanthemum

Replication	Day	Amblyseius cucumeris	Amblyseius barkeri	A. cucumeris + A.barkeri	Control
I	Day 10	3	2	1	7
	Day 17	2	4	2	9
	Day 24	1	2	2	-
	Day 31	-	3	8	38

II	Day 10	2	1	2	9
	Day 17	3	4	1	10
	Day 24	5	2	-	25
	Day 31	3	5	7	34
III	Day 10	1	3	1	8
	Day 17	3	-	1	14
	Day 24	2	1	3	23
	Day 31	5	2	9	32
IV	Day 10	2	4	2	10
	Day 17	1	2	-	11
	Day 24	4	5	5	21
	Day 31	6	6	10	39

(a) How would you estimate the missing values, and perform the analysis of variance?
(b) Is there a significant difference between the results obtained in problem 3 and the present case?

Solution:

a.

When we have missing data, we use the following formula to estimate the value of the missing data:

$$\widehat{M} = \frac{r\,U + b\,(A_i B_j) - (A_i)}{(r-1)(b-1)}$$

where

$\widehat{M}$ = the estimated value of the missing observation

r = Number of replications

U = total for unit containing the missing observation

b = the level of subplot factors

$A_i B_j$ = total of observed values of the treatment combination that contain the missing observation

A_i = total of all observations that receive the ith level of A.

To find the estimated value for the missing observation, we compute the table of totals with the missing observations.

The table of totals computed from the original data are given below.

The Replication x Biological Control (Factor *A*) Totals Computed from Data

	Biological Controls				Rep. Total (*R*)
Replication	*Amblyseius cucumeris*	*Amblyseius barkeri*	*A. cucumeris + A. barkeri*	Control	
I.	6	11	13	54	84
II.	13	12	10	78	113
III.	11	6	14	77	108
IV	13	17	17	81	128
Duration Total (*A*)	43	46	54	290	
Grand Total (*G*)					433

Duration x Biological Control (Factor *B*) Table of Totals

	Biological Controls				Duration Total (*B*)
Duration	*Amblyseius cucumeris*	*Amblyseius barkeri*	*A. cucumeris + A. barkeri*	Control	
Day 10	8	10	6	34	58
Day 17	9	10	4	44	67
Day 24	12	10	10	69	101
Day 31	14	16	34	143	207

Using the data in the above tables, we estimate the value of the missing data as follows:

$$\widehat{M} = \frac{4(54) + 4(290) - (69)}{(4-1)(4-1)} = 22.4 \text{ which is rounded to } 22$$

The estimated value for all other missing observation is shown below:

$$\widehat{M} = \frac{4(6) + 4(43) - (14)}{(4-1)(4-1)} = 4.11 \text{ which is rounded to } 4$$

$$\widehat{M} = \frac{4(10) + 4(54) - (10)}{(4-1)(4-1)} = 2.88 \text{ which is rounded to } 3$$

$$\widehat{M} = \frac{4(6) + 4(46) - (10)}{(4-1)(4-1)} = 2.0$$

$$\widehat{M} = \frac{4(17) + 4(54) - (4)}{(4-1)(4-1)} = 3.33 \text{ which is rounded to } 3$$

With the estimated values placed among the original data, we perform the analysis as follows:

The Replication x Biological Control (Factor *A*) Totals Computed from Data

	Biological Controls				Rep. Total (*R*)
Replication	*Amblyseius cucumeris*	*Amblyseius barkeri*	*A. cucumeris + A. barkeri*	Control	
I.	10	11	13	76	110
II.	13	12	13	78	116
III.	11	8	14	77	110
IV	13	17	20	81	131
Duration Total (*A*)	47	48	60	312	
Grand Total (*G*)					467

<u>Duration x Biological Control (Factor *B*) Table of Totals</u>

Duration	Amblyseius cucumeris	Amblyseius barkeri	A. cucumeris + A. barkeri	Control	Duration Total (*B*)
Day 10	8	10	6	34	58
Day 17	9	12	7	44	72
Day 24	12	10	13	91	126
Day 31	18	16	34	143	211

Since this problem is a modification of the previous problem, the degree of freedom associated with each source of variation remains the same.

The correction factor is:

$$C = \frac{G^2}{rab}$$

$$= \frac{(467)^2}{4 \times 4 \times 4} = 3407.64$$

$$\text{Total } SS = \sum X^2 - C$$

$$= [(3)^2 + (2)^2 + \ldots + (39)^2] - 3407.64$$

$$= 5325.36$$

$$\text{Replication } SS = \frac{\sum R^2}{ab} - C$$

$$= \frac{(110)^2 + (116)^2 + (110)^2 + (131)^2}{16} - 3407.64$$

$$= 18.42$$

$$SS_A \text{ (Biological Controls)} = \frac{\sum A^2}{rb} - C$$

$$= \frac{(47)^2 + (48)^2 + (60)^2 + (312)^2}{16} - 3407.64$$

$$= 3183.42$$

$$\text{Error } SS\,(a) = \frac{\sum (RA)^2}{b} - (C + \text{Rep. } SS + SS_A)$$

$$= \frac{(10)^2 + (11)^2 + \ldots + (81)^2}{4} - (3407.64 + 18.42 + 3183.42)$$

$$= 5.77$$

$$SS_B \text{ (Duration)} = \frac{\sum B^2}{ra} - C$$

$$= \frac{(58)^2 + (72)^2 + (126)^2 + (211)^2}{4 \times 4} - 3407.64$$

$$= 901.42$$

$$SS_{AB} \text{ (Bio logical Controls x Duration)} = \frac{\sum (AB)^2}{r} - (C + SS_A + SS_B)$$

$$= \frac{(8)^2 + (10)^2 + (6)^2 + \ldots + (143)^2}{4 \times 4} - (3407.64 + 3183.42 + 901.42)$$

$$= 1118.77$$

Error $SS(b) = TSS$ - All other sum of squares

$$= 5325.36 - (18.42 + 3183.42 + 5.77 + 901.42 + 1118.77)$$

$$= 97.56$$

Finally, the ANOVA Table presents the completed results

Analysis of Variance of Growth of Broilers Fed Triticale in a Split-Plot Design

Source of Variation	Degree of Freedom	Sum of Squares	Mean Square	F
Replication	3	18.42	4.60	
Biological controls (A)	3	3183.42	1041.23	2379.9**
Error (a)	9	5.77	0.44	
Duration (B)	3	901.42	304.35	103.1**
Biological controls x Duration ($A \times B$)	9	1118.77	123.74	41.9**
Error (b)	36	97.56	2.95	
Total	63	5325.36		

** = significant at 1% level.

b. There is not a significant difference in the results obtained in the present problem and that of the previous problem.

6. In a strip-plot design experiment, researchers were interested more in the interaction between the amount of fertilizer and rainfall than in either of these factors alone. They have

hypothesized that fertilizers produce more range forage in drought than normal years. The data collected from the experiment is shown below.

Range forage yields per acre from fertilizer application in 2 normal precipitation years and 2 drought years.

Rainfall (inches)	Fertilizer Application (lb/acre)	Rep. I	Forage Yield (lb/acre)			
			Rep. II	Rep. III	Rep. IV	Rep. V
	Ammonium Sulfate					
17.5	(300 lb)	4500	4355	4100	4600	4250
12.8	" "	4100	4235	4005	4095	4050
6.0	" "	1400	1325	1200	1375	1390
6.5	" "	1000	1025	995	1020	1000
	Ammonium Nitrate					
17.5	(180 lb)	4300	4235	4025	4165	4120
12.8	" "	4050	4110	4033	4195	4250
6.0	" "	1630	1624	1595	1675	1595
6.5	" "	1060	1028	1000	1029	1016
	16-20-0					
17.5	(375 lb)	4250	4151	4170	4300	4290
12.8	" "	3700	3935	3205	3495	3850
6.0	" "	1400	1302	1296	1315	1390
6.5	" "	1100	1025	991	1022	1019
	Control					
17.5	(None)	3950	3801	3905	3390	3890
12.8	" "	4101	4035	4205	4007	4100
6.0	" "	901	897	906	942	899
6.5	" "	698	674	688	700	645

(a) Perform the analysis of variance.
(b) Compute the coefficient of variation for the problem.
(c) Assume that the observation for the plot receiving ammonium nitrate when rainfall is 12.8 inches in replication II (which is 4110) is missing. How do you estimate this missing value before the analysis is performed?

Solution:
a.

To avoid dealing with large numbers in the computation, each of the original data points has been divided by 100. This should not have any impact on the results.

Rainfall (in)	Fertilizer Application (lb/ac)	Rep. I	Forage Yield (lb/acre)				
			Rep. II	Rep. III	Rep. IV	Rep. V	Total
	Ammonium Sulfate						
17.5	(300 lbs)	45.00	43.55	41.00	46.00	42.5	218.05
12.8	"	41.00	42.35	40.05	40.95	40.5	204.85
6.0	"	14.00	13.25	12.00	13.75	13.9	66.90
6.5	"	10.00	10.25	9.95	10.20	10.0	50.40
Total		110.00	109.40	103.00	110.90	106.9	
	Ammonium Nitrate						
17.5	(180 lb)	43.0	42.35	40.25	41.65	41.20	208.45
12.8	"	40.5	41.10	40.33	41.95	42.50	206.38
6.0	"	16.3	16.24	15.95	16.75	15.95	81.19

		Rep. I	Rep. II	Rep. III	Rep. IV	Rep. V	Total
6.5	"	10.6	10.28	10.00	10.29	10.16	51.33
Total		110.4	109.97	106.53	110.64	109.81	
	16-20-0						
17.5	(375 lb)	42.5	41.51	41.70	43.00	42.90	211.61
12.8	"	37.0	39.35	32.05	34.95	38.50	181.85
6.0	"	14.0	13.02	12.96	13.15	13.90	67.03
6.5	"	11.0	10.25	9.91	10.22	10.19	51.57
Total		104.5	104.13	96.62	101.32	105.49	
	Control						
17.5	(None)	39.50	38.01	39.05	33.90	38.90	189.36
12.8	"	41.01	40.35	42.05	40.07	41.00	204.48
6.0	"	9.01	8.97	9.06	9.42	8.99	45.45
6.5	"	6.98	6.74	6.88	7.00	6.45	34.05
Total		96.50	94.07	97.04	90.39	95.34	

The intermediate tables are calculated from the above.

The Replication x Fertilization (Factor A) Table of Yield Totals

	Yield Total (RA)					Fertilization Total
Fertilization	Rep. I	Rep. II	Rep. III	Rep. IV	Rep. V	(A)
Ammonium Sulfate	110.0	109.40	103.00	110.90	106.90	540.20
Ammonium Nitrate	110.4	109.97	106.53	110.64	109.81	547.35
16-20-0	104.5	104.13	96.62	101.32	105.49	512.06
Control	96.5	94.07	97.04	90.39	95.34	473.34
Rep. Total (R)	421.4	417.57	403.19	413.25	417.54	
Grand Total (G)						2072.95

Replication x Rainfall (B) Table of Yield Totals

	Yield Total (RB)					Rainfall Total
Rainfall	Rep. I	Rep. II	Rep. III	Rep. IV	Rep. V	(B)
17.5	170.00	165.42	162.00	164.55	165.50	827.47
12.8	159.51	163.15	154.48	157.92	162.50	797.56
6.0	53.31	51.48	49.97	53.07	52.74	260.57
6.5	38.58	37.52	36.74	37.71	36.80	187.35

Fertilization x Rainfall (AB) Table of Yield Totals

	Yield Total (AB)			
Fertilization	17.5	12.8	6.0	6.5
Ammonium Sulfate	218.05	204.85	66.90	50.40
Ammonium Nitrate	208.45	206.38	81.19	51.33
16-20-0	211.61	181.85	67.03	51.57
Control	189.36	204.48	45.45	34.05

Determine the degrees of freedom associated with each source of variation. The following sources of variation and the degrees of freedom are identified.

Replication *d.f.* $\qquad = r - 1 \qquad\qquad = 4$

Horizontal factor (A)*d.f.* $\qquad = a - 1 \qquad\qquad = 3$

Error (a) *d.f.* $\qquad = (r - 1)(a - 1) \qquad = 12$

Vertical factor (B) *d.f.* $\qquad = b - 1 \qquad\qquad = 3$

Error (b) *d.f.* $\qquad = (r - 1)(b - 1) \qquad = 12$

$A \times B$ *d.f.* $\qquad = (a - 1)(b - 1) \qquad = 9$

Error (c) $\qquad = (r - 1)(a - 1)(b - 1) \qquad = 36$

Total *d.f.* $\qquad = r\,ab - 1 \qquad\qquad = 79$

The correction factor is:

$$C = \frac{G^2}{rab}$$

$$= 53714.02$$

$$\text{Total } SS = \sum X^2 - C$$

$$= [(45)^2 + (43.55)^2 + \ldots + (6.45)^2] - 353714.02$$

$$= 17915.79$$

$$\text{Replication } SS = \frac{\sum R^2}{ab} - C$$

$$= \frac{(421.4)^2 + (417.57)^2 + (403.19)^2 + (413.25)^2 + (417.54)^2}{16} - 53714.02$$

$$= 12.23$$

$$SS_A \text{ (Fertilizer)} = \frac{\sum A^2}{rb} - C$$

$$= \frac{(540.20)^2 + (547.35)^2 + (512.06)^2 + (473.34)^2}{20} - 53714.02$$

$$= 169.19$$

$$\text{Error } SS(a) = \frac{\sum (RA)^2}{b} - (C + \text{Rep. } SS + SS_A)$$

$$= \frac{(110)^2 + (109.4)^2 + \ldots + (95.34)^2}{4} - (53714.02 + 12.23 + 169.19)$$

$$= 20.63$$

$$SS_B \text{ (Rainfall)} = \frac{\sum B^2}{ra} - C$$

$$= \frac{(827.47)^2 + (797.56)^2 + (260.57)^2 + (187.35)^2}{20} - 53714.02$$

$$= 17476.24$$

$$\text{Error } (b)\ SS = \frac{\sum (RB)^2}{a} - (C + \text{Rep. } SS + SS_B)$$

$$= \frac{(170)^2 + (165.42)^2 + \ldots + (36.8)^2}{4} - (53714.02 + 12.23 + 17476.24)$$

$$= 11.13$$

$$SS_{AB} \text{ (Fertilizer x Rainfall)} = \frac{\sum (AB)^2}{r} - (C + SS_A + SS_B)$$

$$= \frac{(218.05)^2 + (204.85)^2 + \ldots + (34.05)^2}{5} - (53714.02 + 169.19 + 17476.24)$$

$$= 178.72$$

$$\text{Error } SS\ (c) = TSS - \text{All other sum of squares}$$

$$= 17915.79 - (12.23 + 169.19 + 20.63 + 17476.24 + 11.13 + 178.72)$$

$$= 47.65$$

Analysis of Variance of Growth of Broilers Fed Triticale in a Split-Plot Design

Source of Variation	Degree of Freedom	Sum of Squares	Mean Square	F
Replication	4	12.23	3.06	
Fertilizer (A)	3	169.19	56.39	32.8**
Error (a)	12	20.63	1.72	
Rainfall (B)	3	17476.24	5825.41	6282.5**
Error (b)	12	11.13	0.93	
Fertilizer x Rainfall ($A \times B$)	9	178.72	19.86	15.0**
Error (c)	36	47.65	1.33	
Total	79	17915.79		

** = significant at 1% level.

As the researchers were interested in the interaction of fertilizer and rainfall, the results show that such interaction is highly significant. This means that the recommendation developed for one particular fertilizer cannot be applied to other amounts of rainfall tested in this experiment.

b. The coefficient of variation for the 3 sources of error are:

$$cv\,(a) = \frac{\sqrt{\text{Error }(a\,)MS}}{\text{Grand Mean}} \times 100$$

$$= \frac{\sqrt{1.71}}{25.91} \times 100 = 5.04\%$$

$$cv\,(b) = \frac{\sqrt{\text{Error }(b\,)MS}}{\text{Grand Mean}} \times 100$$

$$= \frac{\sqrt{0.93}}{25.91} \times 100 = 3.71\%$$

$$cv\,(c) = \frac{\sqrt{\text{Error }(c\,)MS}}{\text{Grand Mean}} \times 100$$

$$= \frac{\sqrt{1.32}}{25.91} \times 100 = 4.43\%$$

c. To estimate the value of the missing observation, we use the following formula:

$$\widehat{M} = \frac{a\,\{b\,[A_iB_j] - A_i\} + r\,(aH + bV - B) - bv + G}{(a-1)(b-1)(r-1)}$$

where:

$a = 4$
$b = 4$
$r = 5$
$A_iB_j = 165.28$
$A_i = 506.25$
$H = 68.87$
$B = 376.47$
$V = 122.05$
$v = 756.46$
$G = 2031.9$

$$\widehat{M} = \frac{4\{[4(165.28) - 506.25]\} + 5[4(68.87) + 4(122.05) - 376.47] - (4)(756.46) + 2031.9}{(4-1)(4-1)(5-1)}$$

$$= 43.38$$

Once estimated the missing value will be placed in the table of observed values, and the analysis of variance is performed the usual way.

Chapter 6

THREE (OR MORE)-FACTOR EXPERIMENTAL DESIGNS

EXERCISES

1. A crop scientist is interested in using a split-split-plot design experiment where there are 3 factors of interest. He wishes to assign the 4 planting dates (P_1, P_2, P_3, P_4) to the main plot arranged in randomized complete blocks (I, II, III, and IV). Subplots are to be Fumigated (F_1) and nonfumigated (F_2) for mite control. The 4 harvest dates (H_1, H_2, H_3, H_4) are to be assigned to the sub-subplots. Show how you would lay out this experiment in the field.

Solution:

The steps in randomizing the factors in the main plot and subplots are given below.

STEP 1:

P2	P1	P4	P3
P4	P3	P1	P2
P4	P1	P2	P3
P3	P2	P1	P4
Rep I.	Rep. II	Rep. III	Rep. IV

STEP 2:

P2F1 | P2F2	P1F1 | P1F2	P4F1 | P4F2	P3F2 | P3F1
P4F2 | P4F1	P3F2 | P3F1	P1F2 | P1F1	P2F2 | P2F1
P4F1 | P4F2	P1F2 | P1F1	P2F2 | P2F1	P3F1 | P3F2
P3F2 | P3F1	P2F2 | P2F1	P1F1 | P1F2	P4F1 | P4F2
Rep I.	Rep. II	Rep. III	Rep. IV

STEP 3:

Rep I

P2F1H1	P2F2H4
P2F1H4	P2F2H1
P2F1H3	P2F2H3
P2F1H2	P2F2H2

P4F2H4	P4F1H2
P4F2H1	P4F1H3
P4F2H3	P4F1H1
P4F2H2	P4F1H4

P4F1H2	P4F2H4
P4F1H1	P4F2H3
P4F1H4	P4F2H2
P4F1H3	P4F2H1

P3F2H4	P3F1H2
P3F2H1	P3F1H3
P3F2H2	P3F1H1
P3F2H3	P3F1H4

Rep. II

P1F1H2	P1F2H3
P1F1H1	P1F2H4
P1F1H3	P1F2H1
P1F1H4	P1F2H2

P3F2H4	P3F1H1
P3F2H1	P3F1H2
P3F2H3	P3F1H3
P3F2H2	P3F1H4

P1F2H3	P1F1H2
P1F2H4	P1F1H1
P1F2H2	P1F1H4
P1F2H2	P1F1H3

P2F2H3	P2F1H4
P2F2H2	P2F1H1
P2F2H1	P2F1H3
P2F2H4	P2F1H2

Rep. III

P4F1H2	P4F2H3
P4F1H3	P4F2H1
P4F1H2	P4F2H2
P4F1H1	P4F2H4

P1F2H3	P1F1H2
P1F2H4	P1F1H3
P1F2H1	P1F1H1
P1F2H2	P1F1H4

P2F2H1	P2F1H4
P2F2H2	P2F1H1
P2F2H4	P2F1H2
P2F2H3	P2F1H3

P1F1H1	P1F2H2
P1F1H2	P1F2H1
P1F1H3	P1F2H4
P1F1H4	P1F2H3

Rep. IV

P3F2H1	P3F1H2
P3F2H4	P3F1H3
P3F2H2	P3F1H4
P3F2H3	P3F1H1

P2F2H4	P2F1H3
P2F2H3	P2F1H2
P2F2H2	P2F1H1
P2F2H1	P2F1H4

P3F1H2	P3F2H3
P3F1H1	P3F2H2
P3F1H4	P3F2H1
P3F1H3	P3F2H4

P4F1H3	P4F2H1
P4F1H4	P4F2H3
P4F1H2	P4F2H2
P4F1H1	P4F2H4

2. Certain varieties of cotton such as Acala SJ-2 show potassium deficiency even when the soils, by test, do not show any deficiency. This problem is severe on soils with high levels of Verticillium wilt. In a 2 x 2 x 3 split-split-plot design experiment to determine whether potassium deficiency in cotton is disease-induced, the researchers have assigned the fertilizers (NPK) to the main plot, the treatments to the subplots, and the depth soil as a sub-subplot factor. They have collected soil samples from 2 depths in fumigated and nonfumigated plots and recorded the following data.

<u>Effect of soil fumigation on levels of N, P, and K at 2 sampling depths.</u>

	Parts Per Million								
	N			P			K		
Treatment	Rep.1	Rep.2	Rep.3	Rep.1	Rep.2	Rep.3	Rep.1	Rep.2	Rep.3
				0-12 in. depth					
Fumigated	30.2	30.3	31.0	24.2	24.5	23.8	92.0	94.0	93.3
Nonfumigated	68.0	67.8	68.2	19.0	19.3	19.5	104.8	105.1	104.6
				12-24 in. depth					
Fumigated	28.2	30.3	28.5	20.2	21.5	20.8	89.0	88.4	88.7
Nonfumigated	30.2.	29.8	30.1	17.9	18.3	18.5	93.8	93.0	92.9

(a)   Do a main-plot analysis.
(b)   Do a subplot analysis.
(c)   Do a sub-subplot analysis.
(d)   Compute the coefficient of variation.

Solution:
a, b, and c are performed together in the analysis below.

The Replication x Fertilizer (Factor *A*) Table of Yield Totals

Fertilizer	Yield Total (*RA*)			Fertilizer Total (*A*)
	Rep. I	Rep. II	Rep. III	
N	156.6	158.2	157.8	472.6
P	81.3	83.6	82.6	247.5
K	379.6	380.5	379.5	1139.6
Rep. Total (*R*)	617.5	622.3	619.9	
Grand Total (*G*)				1859.7

The degree of freedom for the various sources are:

Replication *d.f.* =	$r - 1$	$= 2$
Main-plot factor (*A*)*d.f.*=	$a - 1$	$= 2$
Error (*a*) *d.f.* =	$(r - 1)(a - 1)$	$= 4$
Subplot factor (*B*) *d.f.* =	$b - 1$	$= 1$
A x *B* *d.f.*=	$(a - 1)(b - 1)$	$= 2$
Error (*b*) *d.f.* =	$a (r - 1)(b - 1)$	$= 6$
Sub-subplot factor (*C*)=	$c - 1$	$= 1$
A x *C* *d.f.*=	$(a - 1)(c - 1)$	$= 1$
B x *C* *d.f.*=	$(b - 1)(c - 1)$	$= 1$
A x *B* x *C* *d.f.*=	$(a - 1)(b - 1)(c - 1)$	$= 3$
Error (*c*)=	$ab (r - 1)(c - 1)$	$= 12$
Total *d.f.* =	$r\,abc - 1$	$= 35$

The correction factor is:

$$C = \frac{G^2}{rabc}$$

$$= \frac{(1859.7)^2}{3 \times 3 \times 2 \times 2} = 96069.00$$

$$\text{Total } SS = \sum X^2 - C$$

$$= [(30.2)^2 + (30.3)^2 + \ldots + (92.9)^2] - 96069.00$$

$$= 39648.63$$

$$\text{Replication } SS = \frac{\sum R^2}{abc} - C$$

$$= \frac{(617.5)^2 + (622.3)^2 + (619.9)^2}{12} - 96069.00$$

$$= 0.96$$

$$SS_A \text{ (Fertilizer)} = \frac{\sum A^2}{rbc} - C$$

$$= \frac{(472.6)^2 + (247.5)^2 + (1139.6)^2}{(3)(2)(2)} - 96069.00$$

$$= 35872.26$$

$$\text{Error } SS\,(a) = \frac{\sum (RA)^2}{bc} - (C + \text{Rep. } SS + SS_A)$$

$$= \frac{(156.6)^2 + (158.2)^2 + \ldots + (379.5)^2}{(2)(2)} - (96069.00 + 0.69 + 35872.26)$$

$$= 0.20$$

Table of Yield Totals for the Fertilizer x Fumigant (factor A x factor B) Computed from original Data

	Yield Total (AB)	
	Treatment	
Fertilizer	Fumigant	Nonfumigant
N	178.5	294.1
P	135.0	112.5
K	545.4	594.2
Treatment Total (B)	858.9	1000.8

Table of Yield Totals for the Replication x Fertilizer x Fumigation (RAB)

		Yield Total (RAB)		
Fertilizer	Treatment	Rep. I	Rep. II	Rep. III
N	Fumigant	58.4	60.6	59.5
	Nonfumigant	98.2	97.6	98.3
P	Fumigant	44.4	46.0	44.6
	Nonfumigant	36.9	37.6	38.0
K	Fumigant	181.0	182.4	182.0
	Nonfumigant	198.6	198.1	197.5

$$SS_B \text{ (Fumigant)} = \frac{\sum B^2}{rac} - C$$

$$= \frac{(858.9)^2 + (1000.8)^2}{(3)(3)(2)} - 96069.00$$

$$= 559.32$$

$$SS_{AB} \text{ (Fertilization x Fumigation)} = \frac{\sum (AB)^2}{rc} - (C + SS_A + SS_B)$$

$$= \frac{(178.5)^2 + (294.1)^2 + \ldots + (594.2)^2}{(3)(2)} - (96069.00 + 35872.26 + 559.32)$$

$$= 749.93$$

$$\text{Error } (b) \, SS = \frac{\sum (RAB)^2}{c} - (C + \text{Rep. } SS + SS_A + \text{Error } (a)SS + SS_B + SS_{AB})$$

$$= \frac{(58.4)^2 + (60.6)^2 + \ldots + (197.5)^2}{2} - (96069.00 + 0.96 + 35872.26 + 0.20 + 559.32 + 794.93)$$

$$= 2.08$$

Summary Table of Totals for Fertilizer x Depth of Soil (factor A x factor C)

| | Yield Total (AC) | |
| | Depth of soil (in.) | |
Fertilizer	12.0	24.0
N	295.5	177.1
P	130.3	117.2
K	593.8	545.8
Depth of Soil Total (C)	1019.6	840.1

Summary Table of Totals for Fumigation x Depth of Soil (factor B x factor C)

| | Yield Total (BC) | |
| | Depth of Soil (in.) | |
Treatment	12.0	24.0
Fumigated	443.3	415.6
Nonfumigated	576.3	424.5

Summary Table of Totals for Fertilizer x Treatment x Depth of Soil (factor A x factor B x factor C)

| | | Yield Total (ABC) | |
| | | Depth of Soil (in.) | |
Fertilizer	Row Spacing	12.0	24.0
N	Fumigated	91.5	87.0
	Nonfumigated	204.0	90.1
P	Fumigated	72.5	62.5
	Nonfumigated	57.8	54.7
K	Fumigated	279.3	266.1
	Nonfumigated	314.5	279.7

$$SS_C \text{ (Depth of Soil)} = \frac{\sum C^2}{rab} - C$$

$$= \frac{(1019.6)^2 + (840.1)^2}{(3)(3)(2)} - 96069.0$$

$$= 859.01$$

$$SS_{AC} = \frac{\sum (AC)^2}{rb} - (C + SS_A + SS_C)$$

$$= \frac{(295.5)^2 + (177.1)^2 + \ldots + (545.8)^2}{(3)(2)} - (96069.0 + 35872.3 + 859.01)$$

$$= 479.51$$

$$SS_{BC} = \frac{\sum (BC)^2}{ra} - (C + SS_B + SS_C)$$

$$= \frac{(443.3)^2 + (415.6)^2 + (576.3)^2 + (424.5)^2}{(3)(3)} - (96069.0 + 559.3 + 859.01)$$

$$= 427.8$$

$$S_{ABC} = \frac{\sum (ABC)^2}{r} - (C + SS_A + SS_B + SS_C + SS_{AB} + SS_{AC} + SS_{BC})$$

$$= \frac{(91.5)^2 + (87.0)^2 + \ldots + (279.7)^2}{3} - (96069.0 + 35872.3 + 895.01 + 559.3 + 794.9 + 479.51 + 427.8)$$

$$= 612.41$$

Error (c) SS = Total SS - All other sum of squares

$$= 39648.63 - (0.96 + 35872.26 + 0.20 + 559.32 + 794.93 + 2.08 + 895.01$$

$$+ 479.51 + 427.80 + 612.41)$$

$$= 4.14$$

Analysis of Variance of a 3-Factor Cotton Experiment in a Split-Split-Plot Design

Source of Variation	Degrees of Freedom	Sum of Squares	Mean Square	F
Replication	2	0.96	0.48	
Fertilizer (*A*)	2	35872.26	17936.13	*a*
Error (*a*)	4	0.20	0.05	
Treatment (*B*)	1	559.32	559.32	1610.85**
AB	2	794.93	397.47	1144.70**
Error (*b*)	6	2.08	0.35	
Soil Depth (*C*)	1	895.01	895.01	2594.22**
AC	2	479.51	239.75	694.94**
BC	1	427.80	427.80	1240.00**
ABC	2	612.41	306.20	887.55**
Error (*c*)	12	4.14	0.35	
Total	35	39648.63		

** = significant at 1% level, ^{ns} = nonsignificant.

a = there is not sufficient degree of freedom for error (a) for a valid test of significance. You should keep in mind that as a general rule if error d.f. is less 6, *F* values should not be computed.

d. The coefficient of variation for each source of error is given below.

$$cv\ (a)\ =\ \frac{\sqrt{\text{Error } (a)MS}}{\text{Grand Mean}}\ \text{x } 100$$

$$=\ \frac{\sqrt{0.05}}{103.32}\ \text{x } 100\ =\ 0.22\%$$

$$cv\ (b)\ =\ \frac{\sqrt{\text{Error } (b)MS}}{\text{Grand Mean}}\ \text{x } 100$$

$$=\ \frac{\sqrt{0.35}}{103.32}\ \text{x } 100\ =\ 0.57\%$$

$$cv\ (c)\ =\ \frac{\sqrt{\text{Error } (c)MS}}{\text{Grand Mean}}\ \text{x } 100$$

$$=\ \frac{\sqrt{0.34}}{103.32}\ \text{x } 100\ =\ 0.58\%$$

3.   An environmental horticulturist is interested in finding out whether (1) stress-adapted landscapes save water, (2)whether irrigation equal to 15% or less reference evapotranspiration (ET_0) can be applied to established shrubs and ground cover without any drought related injury. This is a 3-factor experiment designed to test the effect of 3 irrigation regimes (no irrigation, 12.0 inches, and 24.0 inches of water) and 2 different irrigation methods (drip and furrow) on the growth of shrubs and ground covers such as Xylosma, Oleander, Coton-easter, Juniper, Ice plant, and Hedera. The experiment is strip-split-plot design replicated 3 times. The data collected at the end of a 2-year period is given below.

Growth of shrubs and ground cover as a function of irrigation water received from April to August.

Plantings	Irrigation Method	Inches of Water Applied	Growth in inches			Total ABC
			Rep. I	Rep. II	Rep. III	
Xylosma	Drip	0.0	8.0	8.4	9.5	25.9
"	"	12.0	19.5	20.1	20.2	59.8
"	"	24.0	30.6	31.0	31.4	93.0
"	Furrow	0.0	6.0	5.4	5.8	17.2
"	"	12.0	12.8	16.9	17.4	47.1
"	"	24.0	28.2	27.6	29.4	85.2
Oleander	Drip	0.0	18.0	19.4	19.5	56.9
"	"	12.0	39.5	40.1	40.3	119.9
"	"	24.0	60.6	59.0	61.4	181.0
"	Furrow	0.0	16.0	15.4	15.7	47.1
"	"	12.0	22.8	36.9	37.4	97.1
"	"	24.0	48.2	47.6	49.4	145.2
Coton easter	Drip	0.0	6.0	6.4	6.5	18.9
"	"	12.0	35.5	31.1	30.6	97.2
"	"	24.0	40.6	41.0	41.3	122.9
"	Furrow	0.0	4.0	4.4	4.8	13.2
"	"	12.0	19.8	16.9	18.4	55.1
"	"	24.0	25.2	27.6	29.5	82.3
Juniper	Drip	0.0	12.0	12.4	12.7	37.1
"	"	12.0	10.5	10.1	10.2	30.8
"	"	24.0	20.6	18.0	19.3	57.9
"	Furrow	0.0	10.0	11.1	10.8	31.9
"	"	12.0	9.8	6.9	7.4	24.1
"	"	24.0	13.2	14.8	15.4	43.4
Ice Plant	Drip	0.0	22.0	18.8	19.5	60.3
"	"	12.0	26.5	28.1	27.2	81.8
"	"	24.0	33.6	33.0	32.4	99.0
"	Furrow	0.0	16.0	15.4	15.8	47.2
"	"	12.0	22.5	26.9	25.4	74.8
"	"	24.0	28.8	29.6	29.9	88.3
Hedera	Drip	0.0	16.0	17.4	18.5	51.9
"	"	12.0	20.5	22.1	20.7	63.3
"	"	24.0	40.6	41.0	41.4	123.0
"	Furrow	0.0	12.0	13.2	14.6	39.8
"	"	12.0	15.8	14.9	14.6	45.3
"	"	24.0	28.2	27.6	29.4	85.2

(a) Perform the analysis of variance.
(b) What conclusions can you draw from the analysis?
(c) Compute the coefficient of variation for the factors.

Solution:

a.

In the table above we have a column of totals (ABC) which has been added to the original data. This will be used later in our calculations.

The following sources of variation and the degrees of freedom are identified.

Replication $d.f.$ =	$r - 1$	= 2
Horizontal factor $(A)d.f.$ =	$a - 1$	= 5
Error (a) $d.f.$ =	$(r - 1)(a - 1)$	= 10
Vertical factor (B) $d.f.$ =	$b - 1$	= 1
Error (b) $d.f.$ =	$(r - 1)(b - 1)$	= 2
$A \times B$ $d.f.$=	$(a - 1)(b - 1)$	= 5
Error (c)=	$(r - 1)(a - 1)(b - 1)$	= 10
Subplot factor (C) $d.f.$ =	$c - 1$	= 2
$A \times C$ $d.f.$=	$(a - 1)(c - 1)$	= 10
$B \times C$ $d.f.$=	$(b - 1)(c - 1)$	= 2
$A \times B \times C$ $d.f.$=	$(a - 1)(b - 1)(c - 1)$	= 10
Error (d)=	ab $(r - 1)(c - 1)$	= 48
Total $d.f.$ =	$rabc - 1$	=107

<u>Summary Table of Replication x Plantings (Factor A) Yield Totals</u>

Planting	Yield Total (RA)			Planting Total (A)
	Rep. I	Rep. II	Rep. III	
Xylosma	105.1	109.4	113.7	328.2
Oleander	205.1	218.4	223.7	647.2
Coton easter	131.1	127.4	131.1	389.6
Juniper	76.1	73.3	75.8	225.2
Ice Plant	149.4	151.8	150.2	451.4
Hedera	133.1	136.2	139.2	408.5
Rep. Total (R)	799.9	816.5	833.7	
Grand Total (G)				2450.1

The correction factor is:

$$C = \frac{G^2}{rabc}$$

$$= \frac{(2450.1)^2}{3 \times 6 \times 2 \times 3} = 55583.24$$

$$\text{Total } SS = \sum X^2 - C$$

$$= [(8)^2 + (8.4)^2 + \ldots + (29.4)^2] - 5558.24$$

$$= 17413.29$$

Replication $SS = \dfrac{\sum R^2}{abc} - C$

$$= \dfrac{(799.9)^2 + (816.5)^2 + (833.7)^2}{36} - 55583.24$$

$$= 15.87$$

SS_A (Plantings) $= \dfrac{\sum A^2}{rbc} - C$

$$= \dfrac{(328.2)^2 + (647.2)^2 + \ldots + (408.5)^2}{(3)(2)(3)} - 55583.24$$

$$= 5512.34$$

Error SS $(a) = \dfrac{\sum (RA)^2}{bc} - (C + \text{Rep. } SS + SS_A)$

$$= \dfrac{(105.1)^2 + (109.4)^2 + \ldots + (139.2)^2}{(2)(3)} - (55583.24 + 15.87 + 5512.34)$$

$$= 26.81$$

Summary Table of Yield Totals for the Replication x Irrigation Method (*RB*)

Irrigation Method	Yield Total (*RB*)			Irrigation Total (*B*)
	Rep. I	Rep. II	Rep. III	
Drip	460.6	457.4	462.6	1380.6
Furrow	339.3	359.1	371.1	1069.5

SS_B (Irrigation method) $= \dfrac{\sum B^2}{rac} - C$

$$= \dfrac{(1380.6)^2 + (1069.5)^2}{(3)(6)(2)} - 55583.24$$

$$= 896.14$$

Error (b) $SS = \dfrac{\sum (RB)^2}{ac} - (C + \text{Rep. } SS + SS_B)$

$$= \dfrac{(460.6)^2 + (457.4)^2 + \ldots + (371.1)^2}{(6)(3)} - (55583.24 + 15.87 + 896.14)$$

$$= 13.55$$

Summary Table of Totals for Plantings x Irrigation Method (factor A x factor B)

Irrigation Method	Total (AB)					
	Xylosma	Oleander	Coton ester	Juniper	Ice Plant	Hedera
Drip	178.7	357.8	239.0	125.8	241.1	238.2
Furrow	149.5	289.4	150.6	99.4	210.3	170.3

Summary Table of Totals for the Replication x Plantings x Irrigation Method (RAB)

Plantings	Irrigation Method	Yield Total (RAB)		
		Rep. I	Rep. II	Rep. III
Xylosma	Drip	58.1	59.5	61.1
	Furrow	47.0	49.9	52.6
Oleander	Drip	118.1	118.5	121.2
	Furrow	87.0	99.9	102.5
Coton easter	Drip	82.1	78.5	78.4
	Furrow	49.0	48.9	52.7
Juniper	Drip	43.1	40.5	42.2
	Furrow	33.0	32.8	33.6
Ice Plant	Drip	82.1	79.9	79.1
	Furrow	67.3	71.9	71.1
Hedera	Drip	77.1	80.5	80.6
	Furrow	56.0	55.7	58.6

$$SS_{AB} = \frac{\sum (AB)^2}{rc} - (C + SS_A + SS_B)$$

$$= \frac{(178.7)^2 + (357.8)^2 + \ldots + (170.3)^2}{(3)(3)} - (55583.24 + 5512.34 + 896.14)$$

$$= 192.85$$

$$\text{Error } (c)\, SS = \frac{\sum (RAB)^2}{c} - (C + \text{Rep. } SS + SS_A + \text{Error } (a)\, SS + SS_B + \text{Error } (b)\, SS + SS_{AB})$$

$$= \frac{(58.1)^2 + (59.5)^2 + \ldots + (58.6)^2}{3} - (55583.24 + 15.87 + 5512.34 + 26.81 + 896.14 +$$

$$+ 13.55 + 192.85)$$

$$= 15.68$$

<u>Summary Table of Totals for Plantings x Amount of Water Applied (Factor *A* x Factor *C*)</u>

| | Yield Total (*AC*) | | |
| | Water Applied (in) | | |
Plantings	0.0	12.0	24.0
Xylosma	43.1	106.9	178.2
Oleander	104.0	217.0	326.2
Coton easter	32.1	152.3	205.2
Juniper	69.0	54.9	101.3
Ice Plant	107.5	156.6	187.3
Hedera	91.7	108.6	208.2
Planting Total	447.4	796.3	1206.4

Summary Table of Totals for Irrigation Method x Amount of Water Applied (Factor *B* x Factor *C*)

| | Yield Total (*BC*) | | |
| | Water Applied (in.) | | |
Irrigation Method	0.0	12.0	24.0
Drip	251	452.8	676.8
Furrow	196.4	343.5	529.6

$$SS_C \text{ (Amount of Water Applied)} = \frac{\sum C^2}{rab} - C$$

$$= \frac{(447.4)^2 + (796.3)^2 + (1206.4)^2}{(3)(6)(2)} - 55583.24$$

$$= 8018.47$$

$$SS_{AC} = \frac{\sum (AC)^2}{rb} - (C + SS_A + SS_C)$$

$$= \frac{(43.1)^2 + (106.9)^2 + \ldots + (208.2)^2}{(3)(2)} - (55583.24 + 5512.34 + 8018.47)$$

$$= 2291.37$$

$$SS_{BC} = \frac{\sum (BC)^2}{ra} - (C + SS_B + SS_C)$$

$$= \frac{(251.0)^2 + (452.8)^2 + \ldots + (529.6)^2}{(3)(6)} - (55583.24 + 896.14 + 8018.47)$$

$$= 120.40$$

$$SS_{ABC} = \frac{\sum (ABC)^2}{r} - (C + SS_A + SS_B + SS_C + SS_{AB} + SS_{AC} + SS_{BC})$$

$$= \frac{(25.9)^2 + (59.8)^2 + \ldots + (85.2)^2}{3} -$$

$$(55583.24 + 5512.34 + 896.14 + 8018.47 + 192.85 + 2291.37 + 120.40)$$

$$= 151.47$$

Error (d) SS = Total SS - All other sum of squares

$$= 17413.29 - (15.87 + 5512.34 + 26.81 + 896.14 + 13.55 + 192.85 + 15.68 +$$

$$8018.47 + 2291.37 + 120.40 + 151.47)$$

$$= 158.35$$

Analysis of Variance of Growth of Shrubs and Ground Cover as a Function of Irrigation water in Strip-Split-Plot Design

Source of Variation	Degree of Freedom	Sum of Squares	Mean Square	F
Replication	2	15.87	7.94	
Plantings (A)	5	5512.34	1102.47	411.36**
Error (a)	10	26.81	2.68	
Irrigation Method (B)	1	896.14	896.14	a
Error (b)	2	13.55	6.78	
AB	5	192.85	38.57	24.72**
Error (c)	10	15.68	1.56	
Water applied (C)	2	8018.47	4009.23	282.48**
AC	10	2291.37	229.13	69.64**
BC	2	120.40	60.20	18.29**
ABC	10	151.47	15.14	4.60**
Error (d)	48	158.35	3.29	
Total	107	17413.29		

** = significant at 1% level. a = Inadequate error degree of freedom for a valid test of significance.

b. The results show that all interactions are highly significant. For appropriate interpretation of large interactions, the analysis should focus on the nature of interactions.

c. The coefficient of variation for each source are:

$$cv\ (a) = \frac{\sqrt{Error\ (a)MS}}{Grand\ Mean} \times 100$$

$$= \frac{\sqrt{2.68}}{22.89} \times 100 = 7.16\%$$

$$cv\,(c) = \frac{\sqrt{\text{Error }(c)\,MS}}{\text{Grand Mean}} \times 100$$

$$= \frac{\sqrt{1.56}}{22.89} \times 100 = 5.46\%$$

$$cv\,(d) = \frac{\sqrt{\text{Error }(d)\,MS}}{\text{Grand Mean}} \times 100$$

$$= \frac{\sqrt{3.29}}{22.89} \times 100 = 7.90\%$$

All the coefficient of variations are relatively small indicating the precision with which each factor has been analyzed. We did not compute the cv(b) as the error (b) degrees of freedom is inadquate for a valid test of significance.

4. Irrigation specialists have found that subsurface drip irrigation has a number of potential advantages over conventional surface irrigation. However, the depth at which a subsurface irrigation system is placed has implications for seed germination. The researcher is interested in the depth of burial of the subsurface drip tape, the depth of seed placement, and irrigation frequency reference evapotranspiration (ET_0). Using a strip-split-plot design experiment with melon seeds, the following data were collected when the experiment was replicated 4 times:

Number of emerged seedlings out of 100 seeds planted 14 days after initial irrigation

Planting Depth (in.)	Burial Depth of Irrigation Tape	Amount of Water Applied	Number of Emerged Seedlings				Total (ABC)
			Rep. I	Rep. II	Rep. III	Rep. IV	
0.5	6	0.25 ET_0	35	32	34	36	137.0
1.5	"	"	53	54	52	56	215.0
2.5	"	"	1	0	0	2	3.0
0.5	9	"	40	38	41	36	155.0
1.5	"	"	38	37	35	38	148.0
2.5	"	"	0	0	1	1	2.0
0.5	12	"	9	10	12	8	39.0
1.5	"	"	30	32	28	30	120.0
2.5	"	"	0	0	0	0	0.0
0.5	6	0.50 ET_0	58	60	62	58	238.0
1.5	"	"	45	47	48	52	192.0
2.5	"	"	6	4	5	3	18.0
0.5	9	"	40	38	42	36	156.0
1.5	"	"	58	57	55	58	228.0
2.5	"	"	3	1	5	4	13.0
0.5	12	"	30	30	32	28	120.0
1.5	"	"	60	52	58	60	230.0
2.5	"	"	5	6	3	4	18.0
0.5	6	0.75 ET_0	50	50	52	46	198.0

Planting Depth			Rep I	Rep II	Rep III	Rep IV	Total (ABC)
1.5	"	"	55	57	58	52	222.0
2.5	"	"	1	1	2	2	6.0
0.5	9	"	48	48	44	46	186.0
1.5	"	"	45	47	42	40	174.0
2.5	"	"	4	3	1	2	10.0
0.5	12	"	28	30	31	26	115.0
1.5	"	"	36	34	38	32	140.0
2.5	"	"	2	2	4	2	10.0

(a)   What do you conclude from the analysis?
(b)   What do the coefficients of variation in this problem show?

Solution:

a. In the above table we have added a column of total values for (ABC). This will be used later in the calculations.

The following sources of variation and the degrees of freedom are identified.

Replication $d.f.$ =	$r - 1$	= 3
Horizontal factor $(A) d.f.$ =	$a - 1$	= 2
Error (a) $d.f.$ =	$(r - 1)(a - 1)$	= 6
Vertical factor (B) $d.f.$ =	$b - 1$	= 2
Error (b) $d.f.$ =	$(r - 1)(b - 1)$	= 6
$A \times B$ $d.f.$=	$(a - 1)(b - 1)$	= 4
Error (c)=	$(r - 1)(a - 1)(b - 1)$	= 12
Subplot factor (C) $d.f.$ =	$c - 1$	= 2
$A \times C$ $d.f.$=	$(a - 1)(c - 1)$	= 4
$B \times C$ $d.f.$=	$(b - 1)(c - 1)$	= 4
$A \times B \times C$ $d.f.$=	$(a - 1)(b - 1)(c - 1)$	= 8
Error (d)=	$ab (r - 1)(c - 1)$	= 54
Total $d.f.$ =	$rabc - 1$	=107

<u>Summary Table of Replication x Planting Depth (Factor A) Yield Totals</u>

	Yield Total (RA)				Planting Total (A)
Planting Depth (in)	Rep. I	Rep. II	Rep. III	Rep. IV	
0.5	338	336	350	320	1344
1.5	420	417	414	418	1669
2.5	22	17	21	20	80
Rep. Total (R)	780	770	785	758	
Grand Total (G)					3093

The correction factor is:

$$C = \frac{G^2}{rabc}$$

$$= \frac{(3093)^2}{4 \times 3 \times 3 \times 3} = 88580.1$$

$$\text{Total } SS = \sum X^2 - C$$

$$= [(35)^2 + (32)^2 + \ldots + (2)^2] - 88580.1$$

$$= 49540.92$$

$$\text{Replication } SS = \frac{\sum R^2}{abc} - C$$

$$= \frac{(780)^2 + (770)^2 + (785)^2 + (758)^2}{27} - 88580.1$$

$$= 15.81$$

$$SS_A \text{ (Planting Depth)} = \frac{\sum A^2}{rbc} - C$$

$$= \frac{(1344)^2 + (1669)^2 + (80)^2}{(4)(3)(3)} - 88580.1$$

$$= 39150.39$$

$$\text{Error } SS\ (a) = \frac{\sum (RA)^2}{bc} - (C + \text{Rep. } SS + SS_A)$$

$$= \frac{(338)^2 + (336)^2 + \ldots + (20)^2}{(3)(3)} - (88580.1 + 15.81 + 39150.39)$$

$$= 38.50$$

Summary Table of Yield Totals for the Replication x Burial Depth (*RB*)

Burial Depth (in)	Yield Total (*RB*)				Burial Depth Total (*B*)
	Rep. I	Rep. II	Rep. III	Rep. IV	
6	304	305	313	307	1229
9	276	269	266	261	1072
12	200	196	206	190	792

$$SS_B \text{ (Burial Depth)} = \frac{\sum B^2}{rac} - C$$

$$= \frac{(1229)^2 + (1072)^2 + (1229)^2}{(4)(3)(3)} - 88580.1$$

$$= 2722.39$$

$$\text{Error } (b) \; SS = \frac{\sum (RB)^2}{ac} - (C + \text{Rep. } SS + SS_B)$$

$$= \frac{(304)^2 + (305)^2 + \ldots + (190)^2}{9} - (88580.1 + 15.81 + 2722.39)$$

Summary Table of Totals for Planting Depth x Burial Depth factor A x factor B)

Planting Depth (in)	Seed Total (AB)		
	0.5	1.5	2.5
6	573	629	27
9	497	550	25
12	274	490	28

Summary Table of Totals for the Replication x Planting Depth x Burial Depth (RAB)

Planting Depth (in.)	Burial Depth of irrigation tape (in.)	Yield Total (RAB)			
		Rep. I	Rep. II	Rep. III	Rep. IV
0.5	6	143	142	148	140
	9	128	124	127	118
	12	67	70	75	62
1.5	6	153	158	158	160
	9	141	141	132	136
	12	126	118	124	122
2.5	6	8	5	7	7
	9	7	4	7	7
	12	7	8	7	6

$$SS_{AB} = \frac{\sum (AB)^2}{rc} - (C + SS_A + SS_B)$$

$$= \frac{(573)^2 + (629)^2 + \ldots + (28)^2}{(4)(3)} - (88580.1 + 39150.39 + 2722.39)$$

$$= 2113.22$$

$$\text{Error } (c) \, SS = \frac{\sum (RAB)^2}{c} - (C + \text{Rep. } SS + SS_A + \text{Error } (a) \, SS + SS_B + \text{Error } (b) \, SS + SS_{AB})$$

$$= \frac{(143)^2 + (143)^2 + \ldots + (6)^2}{3} - (88580.1 + 15.81 + 39150.39 + 38.5 + 2722.39 +$$

$$+ 17.83 + 2113.22)$$

$$= 33.44$$

Summary Table of Totals for Factor A x Factor C

Planting Depth (in.)	Yield Total (AC)		
	0.25 ETo	0.50 ETo	0.75 ETo
0.5	331	514	499
1.5	483	650	536
2.5	5	49	26
Water Applied Total (C)	819	1213	1061

Summary Table of Totals for Irrigation Tape Depth Factor B x Factor C (Amount of Water Applied)

Burial Depth (in.)	Yield Total (BC)		
	0.25 ETo	0.50 ETo	0.75 ETo
6	355	448	426
9	305	397	370
12	159	368	265

$$SS_C \text{ (Amount of Water Applied)} = \frac{\sum C^2}{rab} - C$$

$$= \frac{(819)^2 + (1213)^2 + (1061)^2}{(4)(3)(3)} - 88580.1$$

$$= 2193.56$$

$$SS_{AC} = \frac{\sum (AC)^2}{rb} - (C + SS_A + SS_C)$$

$$= \frac{(331)^2 + (514)^2 + \ldots + (26)^2}{(4)(3)} - (88580.1 + 39150.39 + 2193.56)$$

$$= 821.39$$

$$SS_{BC} = \frac{\sum (BC)^2}{ra} - (C + SS_B + SS_C)$$

$$= \frac{(355)^2 + (448)^2 + \ldots + (265)^2}{(4)(3)} - (88580.1 + 2722.39 + 2193.56)$$

$$= 393.06$$

$$SS_{ABC} = \frac{\sum (ABC)^2}{r} - (C + SS_A + SS_B + SS_C + SS_{AB} + SS_{AC} + SS_{BC})$$

$$= \frac{(137)^2 + (215)^2 + \ldots + (10)^2}{4} - (88580.1 + 39150.39 + 2722.39 + 2193.56 + 2113.22 + 821.39 + 393.06)$$

$$= 1831.67$$

Error (d) SS = Total SS - All other sum of squares

$$= 49540.92 - (15.81 + 39150.39 + 38.5 + 2722.39 + 17.83 + 2113.22 + 33.44 +$$

$$2193.56 + 821.39 + 393.06 + 1831.67)$$

$$= 209.67$$

Analysis of Variance of Subsurface Drip Irrigation in Strip-Split-Plot Design

Source of Variation	Degree of Freedom	Sum of Squares	Mean Square	F
Replication	3	15.81	5.27	
Planting depth (A)	2	39150.39	19575.19	3050.68**
Error (a)	6	38.5	6.42	
Tape burial depth (B)	2	2722.39	1361.19	457.97**
Error (b)	6	17.83	2.97	
AB	4	2113.22	528.31	189.56**
Error (c)	12	33.44	2.79	
Water applied (C)	2	2193.56	1096.78	282.48**
AC	4	821.39	205.34	52.89**
BC	4	393.06	98.26	25.31**
ABC	8	1831.67	228.96	58.97**
Error (d)	54	209.67	3.88	
Total	107	49540.92		

** = significant at 1% level.

The results show that all factors including their interactions are highly significant.

b. The coefficient of variation for each source are:

$$cv\,(a) = \frac{\sqrt{\text{Error}\,(a)MS}}{\text{Grand Mean}} \times 100$$

$$= \frac{\sqrt{6.42}}{28.64} \times 100 = 8.83\%$$

$$cv(b) = \frac{\sqrt{\text{Error }(b)\,MS}}{\text{Grand Mean}} \times 100$$

$$= \frac{\sqrt{2.97}}{28.64} \times 100 = 6.00\%$$

$$cv(c) = \frac{\sqrt{\text{Error }(c)\,MS}}{\text{Grand Mean}} \times 100$$

$$= \frac{\sqrt{2.79}}{28.64} \times 100 = 5.83\%$$

$$cv(d) = \frac{\sqrt{\text{Error }(d)\,MS}}{\text{Grand Mean}} \times 100$$

$$= \frac{\sqrt{388}}{28.64} \times 100 = 6.84\%$$

All the coefficient of variations are relatively small indicating the precision with which each factor has been analyzed.

5. In a fractional factorial design experiment (half-replicate), a crop scientist is interested in knowing the impact of five different fertilizers each at 2 levels on yield of corn. The scientist wishes to have 2 blocks in each of the 2 replications. Show how you would lay out the experiment for the scientist.

From Appendix G, we select Plan 3 which meets the requirement for a 2^5 factoria experiment to be conducted in a half-replicate.

The selected plan has the following treatment numbers assigned to each block as shown below.

A Half-Replicate of a 2^5 Factorial

Treatment No.	1	2	3	4
	5	6	7	8
	9	10	11	12
	13	14	15	16
Block I	(1)	ac	ae	ad
	ab	bc	be	bd
	acde	de	cd	ce
Block II	bcde	abde	abcd	abce

Randomly assign the block arrangement in the basic plan to the blocks in the field. We have selected the following arrangements.

Block 1	Block 2
1 (1)	1 ac
2 ae	2 ad
3 ab	3 bc
4 be	4 bd
5 acde	5 de
6 cd	6 ce
7 bcde	7 abde
8 abcd	8 abce

Block 1	Block 2
1 (1)	1 ac
2 ae	2 ad
3 ab	3 bc
4 be	4 bd
5 acde	5 de
6 cd	6 ce
7 bcde	7 abde
8 abcd	8 abce

Replication I   Replication II

6. Suppose that in the previous example, the data collected from the experiment are as shown below.

Corn yield from a 2^5 factorial experiment conducted as a half-replicate with 2 blocks each containing 2 replicates.

	Grain Yield (bu/ac) Block 1			Grain Yield (bu/ac) Block 2	
Treatment	Rep. I	Rep. II	Treatment	Rep. I	Rep. II
(1)	132	125	ac	112	114
ae	151	145	ad	122	132
ab	136	138	bc	145	148
be	144	142	bd	161	158
acde	171	162	de	145	138
cd	154	159	ce	162	168
bcde	143	138	abde	144	140
abcd	132	136	abce	155	159

(a) Perform the analysis of variance.
(b) Which one of the main effects and interactions were highly significant?

Solution:

a.

Table of Corn yield from a 2^5 factorial experiment conducted as a half replicate with 2 blocks each containing 2 replictes

	Grain Yield (bu/ac) Block 1				Grain Yield (bu/ac) Block 2		
Treatment	Rep. I	Rep.II	Total	Treatment	Rep. I	Rep. II	Total
(1)	132	125	257	*ac*	112	114	226
ae	151	145	296	*ad*	122	132	254
ab	136	138	274	*bc*	145	148	293
be	144	142	286	*bd*	161	158	319
acde	171	162	333	*de*	145	138	283
cd	154	159	313	*ce*	162	168	330
bcde	143	138	281	*abde*	144	140	284
abcd	132	136	268	*abce*	155	159	314
Total (*RB*)	1163	1145			1146	1157	

Go to appendix G and select plan 3 which accomodates a 25 factorial in a half replicate. We have identified the following effects and the d.f. for this problem

Effects	d.f.
Rep.	1
Block	1
Rep.xBlock	1
Main	5
2-factor	5
3-factor	3
Error	15
Total	31

Calculate the replication total (R), block total (B), rep x block total (RB) as follows:

$$\text{Rep. I } (R_1) = 1163 + 1146 = 2309$$

$$\text{Rep. II } (R_2) = 1145 + 1157 = 2302$$

$$\text{Block 1 } (B_1) = 1163 + 1145 = 2308$$

$$\text{Block 2 } (B_2) = 1146 + 1157 = 2303$$

$$G = 2309 + 2302 = 4611$$

The correction factor is:

$$C = \frac{G^2}{rt}$$

$$C = \frac{(4611)^2}{2 \times 16} = 664416.3$$

$$\text{Total } SS = \sum X^2 - C$$

$$= [(132)^2 + (125)^2 + \ldots + (159)^2] - 664416.30$$

$$= 6634.72$$

$$\text{Replication } SS = \frac{\sum R^2}{t} - C$$

$$= \frac{(2309)^2 + (2302)^2}{16} - 664416.3$$

$$= 1.53$$

$$\text{Block } SS = \frac{\sum B^2}{t} - C$$

$$= \frac{(2308)^2 + (2303)^2}{16} - 664416.3$$

$$= 0.78$$

$$\text{Block x Rep. } SS = \frac{\sum (RB)^2}{t/b} - (C + \text{Rep. } SS + \text{Block } SS)$$

$$= \frac{(1163)^2 + (1145)^2 + 1146)^2 + (1157)^2}{16/2} - (664416.3 + 1.53 + 0.78)$$

$$= 26.28$$

Calculate the t factorials as:
Yates' Method of computing

Treatment Combination	t_0	t_1	t_2	t_3	t_4	Contrast	Alias
(1)	257	553	1169	2206	4611	(G)	(G)
a(e)	296	616	1037	2405	165	A	A
b(e)	286	557	1239	102	-67	B	B
c(e)	330	480	1166	63	-45	C	C
d(e)	283	612	83	-14	-205	D	D
ab	274	627	19	-53	-55	AB	AB
ac	226	617	27	42	-223	AC	AC
ad	254	549	36	-87	-5	AD	AD
bc	293	39	63	-132	199	BC	BC
bd	319	44	-77	-73	-39	BD	BD
cd	313	-9	15	-64	-39	CD	CD
abc(e)	314	28	-68	9	-129	ABC	ABC
abd(e)	284	26	5	-140	59	ABD	ABD
acd(e)	333	1	37	-83	73	ACD	ACD
bcd(e)	281	49	-25	32	57	BCD	BCD
abcd	268	-13	-62	-37	-69	ABCD	E

Calcualte the sum of square for the main effect as well as the 2-factor and 3-factor interactions.

Main effect $SS = \dfrac{(A)^2+(B)^2+(C)^2+(D)^2+(E)^2+(F)^2}{(r)(2)^k}$

$= \left[(165)^2+(-67)^2+(-45)^2+(-205)^2+(-69)^2\right]/(2)(16)$

$= 2516.41$

2-factor interaction $SS = \left[(AB)^2+(AC)^2+\ldots+(BF)^2+(AF)^2\right]/(r)(2)^k$

$= \left[(-55)^2+(-223)^2+\ldots+(-39)^2+(-39)^2\right]/(2)(16)$

$= 2981.90$

3-factor interaction $SS = \left[(ABD)^2+(ACD)^2+\ldots+(BDE)^2+(CDE)^2\right]/(r)(2)^k$

$= \left[(-129)^2+(59)^2+(73)^2+(57)^2\right]/(2)(16)$

$= 896.88$

Error SS = Total SS - (the sum of all other SS)

$= 6634.69 - (1.53 + 0.78 + 26.28 + 2516.41 + 2981.9 + 896.88)$

$= 210.91$

Anova Table for Corn Yield

Source of Variation	Degree of Freedom	Sum of Square	Mean Square	F
Replication	1	1.53	1.53	0.10*ns*
Block	1	0.78	0.78	0.06*ns*
Block x Rep.	1	26.28	26.28	1.87*ns*
Main Effect	5	2516.41	503.28	35.79**
2 factor inter.	5	2981.90	596.38	42.41**
3 factor inter.	3	896.88	298.96	21.26**
Error	15	210.91	14.06	
Total	31	6634.69	214.02	

** = highly significant, *ns* = nonsignificant.

b. the main effect and all interactions are highly significant.

Chapter 7

TREATMENT MEANS COMPARISONS

1. In a completely randomized field experiment, measurements were made to study the response of field-grown cassava (*Manihot esculenta* Crantz) to changes in the application of a fertilizer. The objective of the research was to find out if there were any differences in the amount of the dry matter produced under 5 different fertilizer regimes. The mean yield from the experiment with 4 replications were:

Mean yield of dry matter production (tons/ha) of cassava as a result of 5 different fertilizer applications.

Treatment	Treatment Mean (tons/ha)
Control	2.14
50 kg/ha	2.54
100 kg/ha	2.68
150 kg/ha	3.81
200 kg/ha	4.42

(a) Compute the *LSD* value at 5% and 1% levels of significance.
(b) What conclusions can you draw from the comparisons?

Solution:

a.

Treatment	Treatment Mean (tons/ha)
Control	2.14
50 kg/ha	2.54
100 kg/ha	2.68
150 kg/ha	3.81
200 kg/ha	4.42

In this problem the value for the error mean square was inadvertently left out.

Assume that in this problem, the error mean square is 0.258.

Given that there are 5 treatments with 4 replications, the error degrees of freedom will be 15.

From Appendix I, the t values for the 5% and 1% are:
$t_{.05}=$ 2.131
$t_{.01}=$ 2.947

$$LSD_{.05} = (t_{.05})\sqrt{\frac{2s^2}{r}}$$

$$= 2.179\sqrt{\frac{2(0.258)}{4}} = 0.281$$

$$LSD_{.01} = 3.055\sqrt{\frac{2(.258)}{4}} = 0.394$$

Treatment	Treatment Mean (tons/ha)	Difference from Control
Control	2.14	-
50 kg/ha	2.54	0.40**
100 kg/ha	2.68	0.14ns
150 kg/ha	3.81	1.13**
200 kg/ha	4.42	2.28**

b. It appears from the results that field grown cassva does respond to fertilizer under all conditions except when 100 kg/ha of fertilizer applied.

2. Suppose that in the previous example the treatment means were computed from the 4 replications where some data were lost as a result of a natural disaster, as shown below.

Mean yield of dry matter production (tons/ha) of cassava as a result of 5 different fertilizer applications.

Treatment	Rep. I	Rep. II	Rep. III	Rep. IV	Treatment Mean
Control	2.20	-	2.25	2.01	2.15
50 kg/ha	2.40	2.56	2.66	2.52	2.54
100 kg/ha	-	2.68	2.79	-	2.74
150 kg/ha	3.00	3.56	4.00	4.66	3.81
200 kg/ha	3.50	4.98	5.00	4.20	4.42

(a) What type of a comparison of means would be appropriate?
(b) Compute the *LSD* at the 5% and 1% level of significance
(c) Compare each of the mean differences.

Solution:

a. When we have unequal numbers of replications in an experiment, the appropriate comparison is the planned-pair comparisons.

We need to compute the error mean square so that we can compute the LSD values. The ANOVA table for this problem is:

Source of Variation	Degree of Freedom	Sum of Square	Mean Square
Treatment	4	22.76	5.69
Error	15	3.09	0.21
Total	19	25.86	

From Appendix I determine the t values are:

$t_{.05} = 2.131$
$t_{.01} = 2.947$

$$LSD_{.05} = (t_{.05}) \sqrt{\frac{2s^2}{r}}$$

$$= 2.131 \sqrt{\frac{2(0.206)}{4}} = 0.219$$

$$LSD_{.01} = 2.947 \sqrt{\frac{2(.206)}{4}} = 0.303$$

Treatment	Number of Replications	Mean Yield (tons/ha)	Difference Control	LSD Values 5%	1%
Control	3	2.15	-	-	-
50 kg/ha	4	2.54	0.39	0.22	0.30**
100 kg/ha	2	2.74	0.59	0.09	0.12**
150 kg/ha	4	3.81	1.66	0.22	0.30**
200 kg/ha	4	4.42	2.27	0.22	0.30**

By performing the analysis, we have answered all three questions.

3. In a balanced lattice design experiment where each treatment was replicated 4 times, a natural scientist studied the impact of chlorpyrifos on ash borer infestation. The ash borer represents serious threat to ash and olive trees. The treatments were applied during the moth flight season. The scientist was interested in knowing if there are differences in the number of attacks on the ash tree as a result of the treatments. In performing the analysis of variance, the scientist found the effective error mean square with 16 degrees of freedom to be 1.8.

Treatment Number	Adjusted Mean No. of attack sites
1	3
2	4
3	2
4	3
5	1
6	4
7	2
8	5
9 (Control)	8

(a) Compute the *LSD* for the 5% and 1% levels of significance.
(b) What conclusions can be drawn about the effectiveness of the treatments as compared with the control?

Solution:

a. This experiment was replicated 4 times. Given that the effective error mean square is 1.8 and the degrees of freedom associated with the error mean square is 16, the t values from Appendix I are:

$t_{.05} = \quad 2.120$
$t_{.01} = \quad 2.921$

Thus the LSD values are:

$$LSD_{.05} = (t_{.05}) \sqrt{\frac{2s^2}{r}}$$

$$= 2.120 \sqrt{\frac{2(1.8)}{4}} = 1.908$$

$$LSD_{.01} = 2.921 \sqrt{\frac{2(1.8)}{4}} = 2.628$$

Treatment Number	Adjusted Mean No. of attack sites	Difference from Control
1	3	5**
2	4	4**
3	2	6**
4	3	5**
5	1	7**
6	4	4**
7	2	6**
8	5	3**
9 (Control)	8	-

b. All treatments are highly significant from the control treatment. This implies that there are differences in the number of attacks on the ash tree as a result of the treatments.

4. To make a mean comparison on the data from a 4 x 4 partially balanced triple lattice design experiment, an animal scientist recorded the following data on the role of progesterone in stimulating sexual receptivity in estrogen-treated gilts. The experiment involved 16 ovariectomized gilts who were treated with estradil benzoate (EB), and a control group which did not receive estradil benzoate (treatment number 16). After EB treatment, gilts were moved to an evaluation pen where bores were brought in. Gilts remained in the evaluation pen for 5 minutes, during which time the number of mounts attempted by the bore were recorded. The treatment numbers appear in parenthesis.

Incomplete Block Number	Mounts, Number/5 min.			
	Replication I			
1	7(01)	5(02)	4(03)	4(04)
2	5(05)	2(06)	1(07)	3(08)
3	4(09)	3(10)	4(11)	5(12)
4	3(13)	3(14)	2(15)	1(16)
	Replication II			
1	6(01)	6(05)	6(09)	2(13)
2	4(02)	2(06)	3(10)	3(14)
3	3(03)	3(07)	1(11)	5(15)
4	1(04)	4(08)	3(12)	2(16)
	Replication III			
1	7(01)	5(06)	4(11)	2(16)
2	4(05)	4(02)	3(15)	6(12)
3	1(09)	2(14)	2(03)	4(08)
4	5(13)	3(10)	4(07)	2(04)

(a) How many effective error mean squares are involved in this experiment?
(b) Before performing the *LSD* test, do you need to compute the adjusted

treatment means?

(c)   Compute the mean difference between the control treatment and each of the 15 treatments.

Solution:

a. There are 2 effective error mean squares involved. The first (MS_1) involves treatments that are tested in the same incomplete block, and the second (MS_2) deals with the treatment pairs that never appeared in the same incomplete block.

The error mean square for 2 treatments in the same block:

$$\text{Error } MS = MS \text{ error}\left[1 + (n-1)\mu\right]$$

The error mean square for 2 treatments not in the same block:

$$\text{Error } MS = MS \text{ error } (1 + n\mu)$$

These 2 equations will be used later in our computations.

To determine the values of the effective error mean squares we perform the analysis of variance. The results of this analysis are shown below.

Incomplete Block Number	Mounts, Number/5 min.				Block Total	Cb	Cb^2
	Replication I						
1	7(01)	5(02)	4(03)	4(04)	20	-11	121
2	5(05)	2(06)	1(07)	3(08)	11	10	100
3	4(09)	3(10)	4(11)	5(12)	16	-5	25
4	3(13)	3(14)	2(15)	1(16)	9	6	36
Rep. Total					56	0	
	Replication II						
1	6(01)	6(05)	6(09)	2(13)	20	-4	16
2	4(02)	2(06)	3(10)	3(14)	12	3	9
3	3(03)	3(07)	1(11)	5(15)	12	0	0
4	1(04)	4(08)	3(12)	2(16)	10	7	49
Rep. Total					54	6	
	Replication III						
1	7(01)	5(06)	4(11)	2(16)	18	-11	121
2	4(05)	4(02)	3(15)	6(12)	17	1	1
3	1(09)	2(14)	2(03)	4(08)	9	12	144
4	5(13)	3(10)	4(07)	2(04)	14	-8	64
Rep. Total					58	-6	686
Total (G)	168						

ANOVA Table

Source of Variation	Degree of Freedom	Sum of Square	Mean Square
Replication	2	0.50	0.25
Block (adj.)	9	27.83	3.09
Treatment	15	64.67	4.31
Intrablock error	21	27.00	1.28
Total	47	120.00	

b. We need to compute the adjust treatment means for this problem.

To compute the adjusted treatment means, we use the following formula:

$$\mu_a = \frac{T_a}{k+1}$$

Where:

μ_a = adjusted treatment mean

T_a = adjusted treatment total

$k+1$ = the number of replications (r).

The above equation calls for the adjusted treatment totals, we must convert the treatment totals in the following way :

$$T_a = T + \mu \sum C_b$$

Note that we have to compute μ and the sum of C_b values in order to calculate the T_a.

$$\mu = \frac{\text{Block (adj.) } MS - \text{Intrablock error } MS}{k(r-1)\left[\text{Block (adj.) } MS\right]}$$

$$\mu = \frac{3.093 - 1.286}{4(3-1)(3.093)} = 0.073$$

The sum of the C_b values for all treatments are:

Treatment Number	Sum of C_b
1	-26
2	-7
3	1
4	-12
5	7
6	2
7	2
8	29
9	3
10	-10
11	-16
12	3
13	-6
14	21
15	7
16	2

The sum of C_b for treatment 1 for example is computed as:

$$(-11)+(-4)+(-11) = -26$$

We have computed the values in the above table in the same manner.

Now to compute the T_a for the second treatment, for example, we will have:

$$T_a = 13 + 0.073 \, (-7)$$

$$= 12.48$$

We have computed all the adjusted treatment totals in the same manner and are presented in the following table.

<u>Comparison of the Adjusted Treatment Means with the Control Treatment</u>

Treatment Number[a]	Treatment Total	Adjusted Treatment Total (T_a)	Adjusted Treatment Mean	Difference from Control
1	20	18.10	3.62	2.59*
2	13	12.48	2.49	3.53**
3	9	9.07	1.81	0.78*ns*
4	7	6.12	1.22	0.19*ns*
5	15	15.51	3.10	2.07*ns*
6	9	9.15	1.83	0.80*ns*
7	8	8.15	1.63	0.60*ns*
8	11	13.12	2.62	1.59*ns*
9	11	11.22	2.24	1.21*ns*
10	9	8.27	1.65	0.62*ns*
11	9	7.83	1.57	0.54*ns*
12	14	14.22	2.84	1.81*ns*
13	10	9.56	1.91	0.88*ns*
14	8	9.53	1.90	0.87*ns*
15	10	10.51	2.10	1.07*ns*
16	5	5.15	1.03	-

[a]Those treatment that were tested in the same incomplete block with control are shown in bold.

Having computed the adjusted treatment totals, we are now ready to compute the adjusted treatment mean in the following way:

$$\mu_a = \frac{T_a}{k+1}$$

The adjusted mean for treatment number 1 is given below:

$$\mu_a = \frac{18.10}{4+1} = 3.62$$

The adjusted treatment means for all treatments have been reported in the table above.

Compute the mean difference between the control treatment (treatment 16) and each of the 14 other treatments. This is shown in the 5th column of the above table.

c. We are now ready to make a comparison between the results and the computed *LSD* values. To do so requires computation of the error mean square for treatments in the same block as well as those not in the same block.

$$\text{Error } MS = MS \text{ error}\left[1 + (n - 1)\mu\right]$$

$$\text{Error } MS = 1.286[\,1 + (3 - 1)0.073\,] = 1.47$$

$$\text{Error } MS = MS \text{ error } (1 + n\mu)$$

$$\text{Error } MS = 1.286[\,1 + (3)0.073\,] = 1.57$$

$t_{.05} = 2.080$
$t_{.01} = 2.831$

$$LSD_\alpha = t_\alpha \sqrt{\frac{2\,(\text{effective error } MS_1)}{r}}$$

$$LSD_1 = 2.080 \sqrt{\frac{2\,(1.47)}{3}} = 2.06$$

$$LSD_1 = 2.831 \sqrt{\frac{2\,(1.47)}{3}} = 2.80$$

$$LSD_\alpha = t_\alpha \sqrt{\frac{2\,(\text{effective error } MS_2)}{r}}$$

$$LSD_1 = 2.080 \sqrt{\frac{2\,(1.57)}{3}} = 2.13$$

$$LSD_1 = 2.831 \sqrt{\frac{2\,(1.57)}{3}} = 2.89$$

We compare the values in the 5th column with the *LSD* values to determine their significance.

5. To study the impact of 4 insecticides on the omnivorous looper, a split-plot design experiment was conducted. The insecticide treatments were assigned to the subplots, whereas the concentration rate of active ingredients were assigned to the main plots. The results of the experiment are shown below.

The average number of larvae found per tree 14 days posttreatment from an experiment replicated 3 times

Active Ingredient	Dylox 80 SP	Kryocide 8 F	Lannate L	Orthene 75 SP	Control
1 lb/ac	9.66	10.33	3.00	2.33	17.33
2 lb/ac	7.33	5.00	2.66	2.33	20.00
4 lb/ac	3.66	5.33	2.00	1.33	16.33

(a) Perform a *LSD* test on the data.
(b) What conclusions can be drawn from the analysis?

Solution:

a.

The error mean square(b) for this problem which was 23.5 with 24 error(b) degrees of freedom was inadvertently left out.   The standard error of the mean difference is:

$$S_{\bar{d}} = \sqrt{\frac{2(23.5)}{3}} = 3.96$$

Since there are 24 error degrees of freedom, the table value is 2.064 for 5% and 2.797 for the 1%. Thus, we have:

$$LSD_{\alpha} = (t_{\alpha})(\hat{s}_{d})$$

$$LSD_{.05} = (2.064)(3.96) = 8.17$$

$$LSD_{.01} = (2.797)(3.96) = 11.08$$

b. We can draw conclusions on the basis of whether the result are significant or not as shown below.

The mean difference between different insecticides are given below

	Active Ingredient		
Difference between	1 lb/ac	2 lb/ac	4 lb/ac
Dylox 80 SP & Kryocide 8F	0.67*ns*	2.33*ns*	1.67*ns*
Dylox 80 SP &Lannate L	6.66*ns*	4.67*ns*	1.66*ns*
Dylox 80 SP & Orthene 75SP	7.33*ns*	5.00*ns*	2.33*ns*
Dylox 80 Sp & Control	7.67*ns*	12.67**	12.67**
Kryocide 8 F & Lannate	7.33*ns*	2.34*ns*	3.33*ns*
Kryocide 8F & Orthene 75SP	8.00*ns*	2.67*ns*	4.00*ns*
Kryocide 8F & Control	7.00*ns*	15.00**	11.00*
Lannate L & Orthene 75SP	0.67*ns*	0.33*ns*	0.67*ns*
Lannate L & Control	14.33**	17.34**	14.33**
Orthene & Control	15.00**	17.67**	15.00**

ns = nonsignificant, * = significant at the 5% level, **= significant at the 1% level.

6.   Research has shown that aldicarb is the most preferred pesticide used by potato farmers. In a study conducted in Wisconsin, researchers used a variety of pest management strategies including 6 different chemicals: aldicarb (Temik), phorate (Thimet), disulfoton (Disyston), carbofuran (Furadan), oxamyl (Vydate), and terbufos (Counter) on Atlantic variety potatoes. The data on the yield per acre of potato are given below.

Mean yield of Atlantic variety potatoes treated soil-applied systemic insecticide[a]

Treatment (lb a.i./acre)	Mean yield(cwt per acre)
Temik 15G (3) with fertilizer at planting	327.4
Temik 15G (3) topdress at emergence	369.8
Temik 15G (2) topdress at emergence	360.5
Disyston 15G (3) with fertilizer at planting	318.4
Disyston 15G (3) topdress at emergence	307.4
Disyston 15G (3+3) with fertilizer at planting and topdress at emergence	322.0
Thimet 20G (3) with fertilizer at planting	319.7
Thimet 20G (3) with fertilizer at planting, foliar spray	317.5
Furadan 15G (3) with fertilizer at planting	329.3
Vydate 10G (3) with fertilizer at planting	296.9
Vydate 10G (3) topdress at emergence	315.7
Counter 15G (3) with fertilizer at planting	317.8
Untreated	239.6

[a]Data adapted from Wyman, J. A., Chapman, R. K. and Longridge, J. L. 1982. *Project Report: Vegetable crop entomology field research.* University of Wisconsin, Madison.

(a)   Perform a Duncan's Multiple-Range Test on the data.

Solution:
There were 4 replications in this study. This information was inadvertently left out.

Rank the treatment means in a descending order as shown below:

Rank Order of the Treatments	Mean yield	Rank	Notation
Temik 15G (3) topdress at emergence	369.8	1	A
Temik 15G (2) topdress at emergence	360.5	2	B
Furadan 15G (3) with fertilizer at planting	329.3	3	C
Temik 15G (3) with fertilizer at planting	327.4	4	D
Disyston 15G (3+3) with fertilizer at planting	322.0	5	E
Thimet 20G (3) with fertilizer at planting	319.7	6	F
Disyston 15G (3) with fertilizer at planting	318.4	7	G
Counter 15G (3) with fertilizer at planting	317.8	8	H
Thimet 20G (3) with fertilizer at planting, foliar spray	317.5	9	I
Vydate 10G (3) topdress at emergence	315.7	10	J
Disyston 15G (3) topdress at emergence	307.4	11	K
Vydate 10G (3) with fertilizer at planting	296.9	12	L
Untreated	239.6	13	M

Compute the standard error as:

$$s = \sqrt{\frac{\Sigma x^2}{n} - (\bar{x})^2}$$

$$s = \sqrt{\frac{\Sigma x^2}{n} - (\bar{x})^2} = 29.65$$

Since there are 4 replications in this experiment, the standard error of estimate is:

$$s_{\bar{d}} = \sqrt{\frac{2(29.65)^2}{4}} = 20.96$$

From Appendix H, determine the $r_\alpha(p, f)$ for a = 0.05 and 39 error degrees of freedom. Thus, we have:

$$\boxed{r_{.05} = (2,39) = 2.88}$$
$$\boxed{r_{.05} = (3,39) = 3.02}$$
$$\boxed{r_{.05} = (4,39) = 3.11}$$
$$\boxed{r_{.05} = (5,39) = 3.18}$$
$$\boxed{r_{.05} = (6,39) = 3.23}$$
$$\boxed{r_{.05} = (7,39) = 3.28}$$
$$\boxed{r_{.05} = (8,39) = 3.31}$$
$$\boxed{r_{.05} = (9,39) = 3.34}$$
$$\boxed{r_{.05} = (10,39) = 3.36}$$
$$\boxed{r_{.05} = (11,39) = 3.38}$$
$$\boxed{r_{.05} = (12,39) = 3.39}$$
$$\boxed{r_{.05} = (13,39) = 3.41}$$

The values of the shortest significant range (R_p) are:

$R_2 = \quad 60.365$
$R_3 = \quad 63.299$
$R_4 = \quad 65.186$
$R_5 = \quad 66.653$
$R_6 = \quad 67.701$
$R_7 = \quad 68.749$
$R_8 = \quad 69.378$
$R_9 = \quad 70.006$
$R_{10} = \quad 70.426$
$R_{11} = \quad 70.845$
$R_{12} = \quad 71.054$
$R_{13} = \quad 71.474$

Compare the largest mean with the smallest, using the information provided in the original table of data. If the difference between these means and the R_p value is equal to or greater, the means are significantly different. Repeat this process of comparison for the remaining treatment means as shown below.

B vs M   130.2 >71.47
B vs J   72.9 >71.05
B vs E   62.4 <70.84
A vs M   87.8 >70.43
F vs M   82.4 >70.01
G vs M   80.1 >69.38
D vs M   78.8 >68.75
L vs M   78.2 >67.7
H vs M   77.9 >66.65
K vs M   76.1 >65.19
E vs M   67.8 >63.3
J vs M   57.3 <60.36

We see that there is no significant difference (at the 5% level) between treatment B and E as well as between J and M. Since there is no significant difference between B and E all the other treatments that are a subset of B and E are also nonsignificant at the 5% level.

So, in notation form we have:

B   C   I   A   F   G   D   L   H   K   E   J   M

We see that in the above presentation there is no significant difference (at the 5% level) between treatment B and E, as well as between J and M. So all the treatments that fall between B and E are in the nonsignificant range, and thus it is not necessary to test, for example treatment C with A.

7. In an attempt to provide adequate moisture for germination, cotton farmers in California irrigate their fields before planting. This preseason irrigation has been identified as a major contributor to drainage problems. To remedy this problem a study was conducted to determine if plastic mulch could be used effectively to replace preseason irrigation of cotton fields. Three different plastic mulches (white, black) were used in this experiment on Acala cotton. The experiment was replicated 4 times. Plant height was used as a criterion in determining the differences between the mulches and the control. Perform a grouped mean comparison on the data.

Height (inches) of Acala cotton under difference treatment conditions

Treatment	Rep. I	Rep. II	Rep. III	Rep. IV
Control	10.0	8.5	9.2	11.0
White plastic	20.3	19.5	21.0	19.8
Black Plastic	24.5	23.8	23.6	22.4

(a) Is there a difference in the height of cotton plants from fields with white mulch and black mulch?

(b) Is there a difference between the height of plants receiving white mulch and the control?

(c) Is there a difference between the height of plants receiving black mulch and the control?

Solution:

a.

An analysis of variance was performed and the results are shown below.

ANOVA Table of Height (inches) of Acala cotton under difference treatment conditions

Source of Variation	Degree of Freedom	Sum of Square	Mean Square	F
Replication	3	1.57	0.52	
Treatment	2	419.56	209.77	230.03**
Error	6	5.47	0.91	
Total	11	426.60		

**=significant at 1% level.

A set of orthogonal Contrasts Among Treatments

Contrast	Control 38.7	White Plastic 80.6	Black plastic 94.3	Value of D_i
Control vs W & B	-2	1	1	97.5
White vs Black	0	-1	1	13.7

Compute the sum of squares for the treatment components as:

$$SS_{Control\ vs.\ B\&\ W} = \frac{(97.5)^2}{4(6)} = 396.09$$

$$SS_{White.vsBlack} = \frac{(-13.7)^2}{4(2)} = 23.46$$

Note that the sum of all the mean squares is equal to the treatment sum of squares:

$$MS_{D1} + MS_{D2} = 396.09 + 23.46 = 419.55$$

Perform the F test.

$$F_1 = \frac{396.09}{0.911} = 437.78$$

$$F_2 = \frac{23.46}{0.911} = 25.75$$

When we compare the computed F values with the table F values for the 1% level of significance for the 1 and 6 degrees of freedom (13.74), we conclude that there is
(a). a difference between the height of cotton plants from white mulch and black mulch.
(b). The same conclusion as in (a).
(c). The same conclusion as in (a).

8. To determine whether application of fertilizers produce more range forage in drought than normal years a study was conduct by scientists in Colorado. The results of the experiment produced the following oven dry weight of range forage.

Range forage yield according to rate of nitrogen application in normal and 5 drought years

Replication	Precipitation(in)	Fertilizer Treatment, lb/ac				
		0	30	60	90	120
I.	Normal (16.5)	3520	3600	3750	4150	4390
	Drought (8.5)	2650	3211	3476	4485	4863
	Drought (10.5)	2652	3356	4291	6310	6050
	Drought (6.0)	1432	2397	2513	3200	3252
	Drought (8.0)	1451	2498	2568	2734	3682
II.	Normal (16.5)	3319	3560	3552	4230	4410
	Drought (8.5)	2578	3276	3399	4345	4674
	Drought (10.5)	2602	3416	4300	5810	5980
	Drought (6.0)	1392	2278	2567	3660	3000
	Drought (8.0)	1400	2399	2510	2932	3862
III.	Normal (16.5)	3423	3545	3759	4252	4270
	Drought (8.5)	2760	3200	3498	4375	4542
	Drought (10.5)	2752	3452	4312	6110	6000
	Drought (6.0)	1487	2377	2542	3189	3342
	Drought (8.0)	1493	2456	2575	2831	3732

(a)   Perform a trend or a factorial comparison on the data.

Solution:

To make a trend or factorial comparison, we first perform an analysis of variance. The results of this analysis is shown in the table below.

ANOVA Table of range forage yield according to rate of nitrogen application in normal and 5 drought years

Source of Variation	Degree of Freedom	Sum of Square	Mean Square	F
Replication	2	23747.71	11873.85	
Fertilizer (A)	4	43720625.90	10930156.50	1679.87**
Error (a)	8	52052.56	6506.57	
Precipitation (B)	4	41657559.90	10414390.00	841.22**
Fert. X Precip. (A x B)	16	10116211.00	632263.19	51.07**
Error (b)	40	495203.07	12380.08	
Total	74	96065400.20		

**=Significant at the 1% level.

The figure below shows the treatment means graphed against the level of fertilizer applied. The trend of the mean does not appear to be linear.

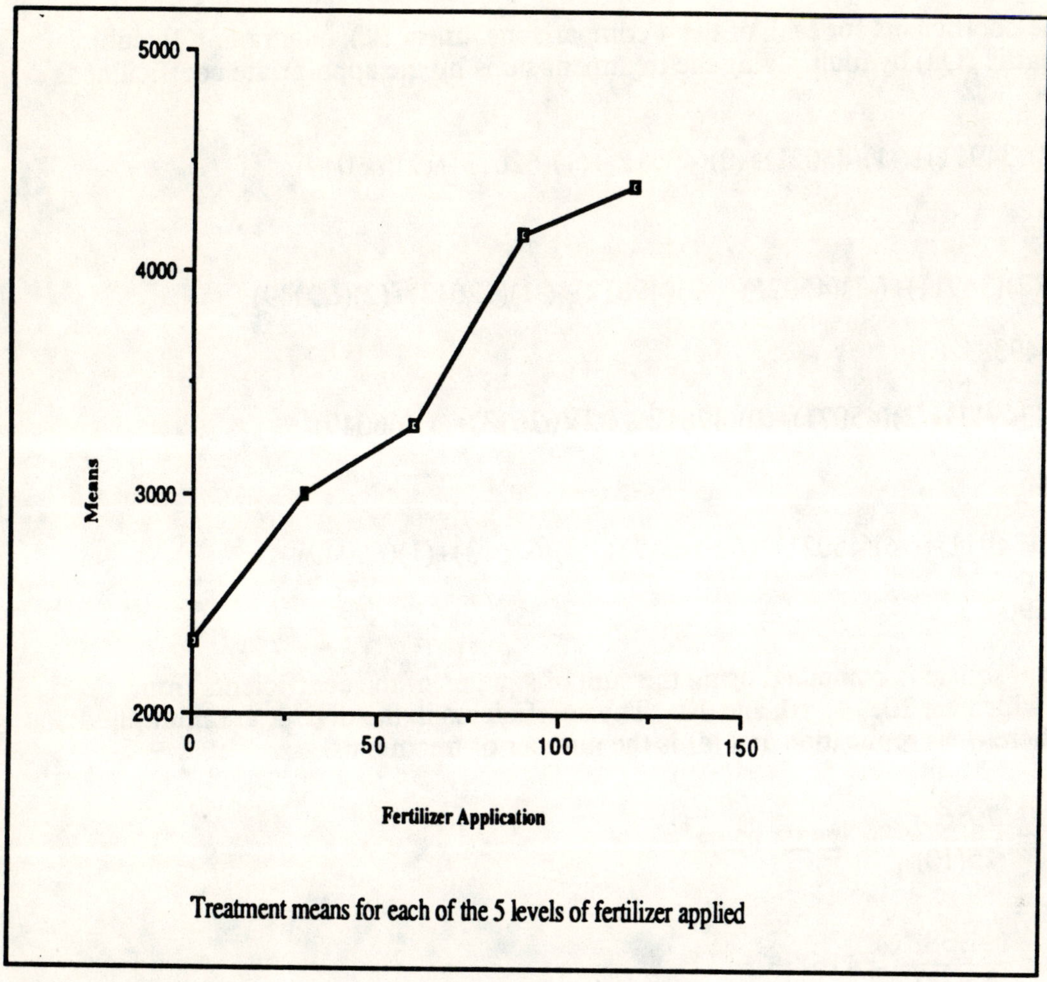

Treatment means for each of the 5 levels of fertilizer applied

The trend of the mean does not appear to be linear.

In order to determine whether the linear component of the trend of the sums is statistically significant and whether the treatment sums (or means) deviate significantly from linearity, use is made of the table of coefficient for orthogonal polynomials given in Appendix J.

Determine the coefficients for the linear, quadratic, and cubic component for $k = 5$ treatments from Appendix J. For the present example the coefficients are:

Coefficients for Linear, Quadratic, Cubic and Quartic Components for $k = 5$ Treatments

	Treatment Totals				
Comparison	34911	45021	49612	62613	66049
Linear	-2	-1	0	1	2
Quadratic	2	-1	-2	-1	2
Cubic	-1	2	0	-2	1
Quartic	1	-4	6	-4	1

Sum the squares for each comparison given in the above Table are given in Appendix J to be 10, 14, 10, and 70.

Compute the coefficients for each of the 4 comparisons, linear (L), quadratic (Q), cubic (C), and Quartic (Qu) by multiplying the treatment sums by the appropriate coefficient as given below:

Linear = (-2)(34911)+(-1)(45021)+(0)(49612)+(1)(62613)+(2)(66049)

Linear =79868

Quadratic= (2)(34911)+(-1)(45021)+(-2)(49612)+(-1)(62613)+(2)(66049)

Quadratic=-4938

Cubic= (-1)(34911)+(2)(45021)+(0)(49612)+(-2)(62613)+(1)(66049)

Cubic= -4046

Quartic= (1)(34911)+(-4)(45021)+(6)(49612)+(-4)(62613)+(1)(66049)

Quartic= -31904

Now the mean square is computed using the sum of squares of the coefficients from Appendix J which are 10, 14, 10, and 70. The coefficients in the divisor are multiplied by $(r)(a)$, where (r) is replication and (a) is the number of treatments.

$$MS_L = SS_L = \frac{(79868)^2}{15(10)} = 42525982.83$$

$$MS_Q = SS_Q = \frac{(-4938)^2}{15(14)} = 116113.54$$

$$MS_C = SS_C = \frac{(-4046)^2}{15(10)} = 109134.11$$

$$MS_{QU} = SS_{QU} = \frac{(-31904)^2}{15(70)} = 969395.44$$

Note that the component sum of square equals the treatment sum of squares.

The final results are presented in the following ANOVA table.

Analysis of Variance of range forage yield partitioned into Linear, Quadratic, and Cubic Components of Response Curve

Source of Variation	Degree of Freedom	Sum of Squares	Mean Square	F	Required F 5%	1%
Replication	2	23747.71	1873.85			
Fertilizer (A)	4	43720625.90	10930156.50	1679.87**	3.84	7.01
Linear	(1)	42525982.80	42525982.80	6535.85**	5.32	11.26
Quadratic	(1)	116113.54	116113.54	17.85**		
Cubic	(1)	109134.11	109134.11	16.77**		
Quartic	(1)	969395.44	969395.44	148.99**		
Error (a)	8	52052.56	6506.57			
Precipitation (B)	4	41657559.90	10414390.00	841.22**		
Fert. X Precip. (A x B)	16	10116211.00	632263.19	51.07**		
Error (b)	40	495203.07	12380.08			
Total	74	139786026.00				

** = Significant at the 1% level.

Chapter 8

SAMPLE DESIGNS OVER TIME

1. Agricultural engineers were interested in evaluating the impact of limited capacity sprinkler irrigation systems on soybean yields. A 3-year field study was conducted using a randomized complete block experimental design. The irrigation treatments included a nonirrigated check (NI), irrigation scheduled by soil moisture depletion (SCH), irrigation which began no earlier than the flowering stage (FL), and irrigation which began no earlier than pod elongation (POD). Soybean yields as influenced by irrigation treatments were recorded as follows.

Soybean yield (bu/acre) as influenced by irrigation treatments

Year	Irrigation Treatment	Rep. I	Rep. II	Rep. III	Rep. IV
1990	NI	37	34	35	35
	POD	50	51	50	49
	FL	53	48	50	51
	SCH	46	48	47	47
1991	NI	32	29	25	30
	POD	34	33	30	35
	FL	43	42	40	38
	SCH	49	48	45	43
1992	NI	30	31	33	32
	POD	40	41	40	38
	FL	43	47	50	45
	SCH	43	47	48	46

(a) Perform a combined analysis over the years.
(b) Perform a test of homogeneity of variance. Can the hypothesis of homogeneous variances be accepted?
(c) What can be said about the nature of the interaction effect between the irrigation treatments and years?

Solution:

a.

To perform the combined analysis, we first need to perform a separate analysis of variance for each year in the experiment. This is done for the years 1990, 1991, and 1992.

The procedure for the analysis for the year 1990 is given below. For the years 1991 and 1992 only the results are presented in the ANOVA Table for the combined years.

$$C = \frac{G^2}{rt}$$

$$= \frac{(731)^2}{16} = 33397.60$$

Total $SS = \displaystyle\sum_{j=1}^{t}\sum_{i=1}^{r} X_{ij}^2 - C$

Total $SS = [(37)^2 + (34)^2 + \text{.......} + (47)^2] - 336397.60$

$\qquad = 631.44$

Replication $SS = \dfrac{\Sigma R_j^2}{t} - C$

$\qquad = \dfrac{(186)^2 + (181)^2 + (182)^2 + (182)^2}{4} - 33397.60$

$\qquad = 3.69$

Treatment $SS = \dfrac{\Sigma T_i^2}{r} - C$

$\qquad = \dfrac{(141)^2 + (200)^2 + (202)^2 + (188)^2}{4} - 33397.60$

$\qquad = 609.69$

Error SS = Total SS - Rep. SS - Treatment SS

$\qquad = 631.44 - 3.69 - 609.69$
$\qquad = 18.06$

Results of the Analyses of Variance for the Individual Years Using an RCB Design with 4 Replications and 4 Irrigation Schemes

Source of Variation	Degree of Freedom	Sum of Squares	Mean Square	F
		1990		
Replication	3	3.69	1.23	0.61ns
Treatment	3	609.69	203.23	101.11**
Error	9	18.06	2.01	
		1991		
Replication	3	45.00	15.00	4.15*
Hybrid	3	717.50	239.17	66.25**
Error	9	32.50	3.61	
		1992		
Replication	3	31.25	10.42	4.87*
Hybrid	3	577.25	192.42	89.92**
Error	9	19.25	2.14	

** = significant at 1% level.

Determine the degrees of freedom associated with each source of variation. Since in the present case year (Y) is considered a random variable, the error term is represented by the year x irrigation interaction MS. Thus, we have:

$$\text{Year } (Y) \; d.f. \qquad = \; y - 1 \qquad\qquad = \; 2$$

$$\text{Reps. within year} = \; y \, (r - 1) \qquad\;\; = \; 9$$

$$\text{Irrigation} \qquad\quad = \; t - 1 \qquad\qquad = \; 3$$

$$\text{Year x Irrigation} = \; (y - 1)(t - 1) \quad = \; 6$$

$$\text{Pooled error} \qquad = \; y \, (r - 1)(t - 1) = \; 27$$

$$\text{Total } d.f. \qquad\quad = \; yrt - 1 \qquad\quad = \; 47$$

Compute the sum of squares for the replications within the year and the pooled error as follows:

$$\text{Rep. within year } SS \; = \; \sum_{i=1}^{y} (\text{Rep. } SS\,)_i$$

$$\text{Pooled Error } SS \; = \; \sum_{i=1}^{y} (\text{Error } SS\,)_i$$

Thus, we have:

$$\text{Reps. within year } SS = 3.69 + 45.00 + 31.25 = 79.94$$

$$\text{Pooled error } SS \; = 18.06 + 32.50 + 19.25 = 69.81$$

Calculate the correction factor as:

$$C \; = \; \frac{\left(\sum\limits_{i=1}^{y} G_i \right)^2}{yrt}$$

$$C \; = \; \frac{(731 + 596 + 654)^2}{48} = 81757.50$$

Compute the sum of squares for the year, treatment, and the year x treatment interaction as follows:

$$\text{Year } SS \; = \; \sum_{i=1}^{y} \frac{G_i^{\,2}}{rt} - C$$

$$\text{Year } SS = \frac{(731)^2 + (596)^2 + (654)^2}{4(4)} - 81757.50$$

$$= 573.29$$

$$\text{Treatment (Irrigation) } SS = \sum_{j=1}^{t} \frac{T_j^2}{yr} - C$$

$$= \frac{\left[(383)^2 + (491)^2 + (550)^2 + (557)^2\right]}{3(4)} - 81575.50$$

$$= 1619.06$$

$$\text{Year x Treatment } SS = \sum_{i=1}^{y} \sum_{j=1}^{t} \frac{(YT)_{ij}^2}{r} - (C + \text{Year } SS - \text{Treatment } SS)$$

Where:

$$(YT)_{ij} = \text{the total of the } j \text{ th treatment in the } i \text{ th year.}$$

$$\text{Year x Treatment } SS = \frac{\left[(141)^2 + (200)^2 + \ldots + (184)^2\right]}{4} - (81575.50 + 573.29 + 1619.06)$$

$$= 285.38$$

Combined ANOVA Table for a RCB Experiment over 3 years

Source of Variation	Degree of Freedom	Sum of Squares	Mean Square	F
Year	2	573.29	286.64	110.67**
Reps. within yr.	9	79.94	8.88	3.43**
Treatment	3	1619.06	539.69	208.37**
Year x Treatment	6	285.38	47.56	18.36**
Pooled Error	27	69.81	2.59	
Total	47	2627.48		

** = Significant at the 1% level.

The results show that all effects and interaction are significant.

b. To perform a test of homogeneity of variance, we need to compute the variance to perform the Bartlett's test. The variance is computed as:

$$s^2 = \frac{\Sigma(x - \bar{x})^2}{n - 1}$$

Year	Treatment	Rep. I	Rep. II	Rep. III	Rep. IV	Mean	Sum of the difference	s_i^2	$\ln s_i^2$
1990	NI	37	34	35	35	35.25	4.75	1.58	0.46
	POD	50	51	50	49	50.00	2.00	0.67	-0.40
	FL	53	48	50	51	50.50	13.00	4.33	1.47
	SCH	46	48	47	47	47.00	2.00	0.67	-0.40
1991	NI	32	29	25	30	29.00	26.00	8.67	2.16
	POD	34	33	30	35	33.00	14.00	4.67	1.54
	FL	43	42	40	38	40.75	14.75	4.92	1.59
	SCH	49	48	45	43	46.25	22.75	7.58	2.03
1992	NI	30	31	33	32	31.50	5.00	1.67	0.51
	POD	40	41	40	38	39.75	4.75	1.58	0.46
	FL	43	47	50	45	46.25	26.75	8.92	2.19
	SCH	43	47	48	46	46.00	14.00	4.67	1.54
Total								49.90	13.15

$$s_i^2 = \frac{\Sigma s_i^2}{k} = \frac{49.90}{12} = 4.16$$

$$\chi^2 = \frac{M}{C}$$

$$M = \upsilon(k \ln s^2 - \Sigma \ln s_i^2)$$

$$C = 1 + \frac{k+1}{3(k)(\upsilon)}$$

To compute M we have $\upsilon = 3$ degrees of freedom per variance. Thus we have:

$$M = (3)[12(1.43)\text{-}13.15]$$

$$=12.03$$

and the value of C is:

$$C = 1+[(12+1)/3(12)(3)]$$

$$= 1.12$$

$$\chi^2 = \frac{12.03}{1.12} = 10.74$$

Since the table value at the .05% level of significance with 11 d.f. is 19.68, we cannot reject the null hypothesis and conclude that the variance are homogeneous.

c. As seen in the ANOVA table, the interaction effect is highly significant.

2. As more attention is being paid to the issue of soil conservation, a number of
practices are suggested as viable methods. Conventional and no-tillage systems are 2
being tested growing different crops. In a study in Nebraska, the performances of
commercial corn hybrids were investigated using the conventional (CT) and no-till (NT)
systems. The other objective of the study was to determine if hybrid x tillage system
interactions exits for grain yield. The experimental design was a randomized complete
block in a split-plot arrangement with 4 replications. The data collected are shown
below.

Grain yield (bu/acre) under conventional and no-tillage systems for 7 corn hybrids

| | | CT | | | | NT | | | |
| | | Replications | | | | Replications | | | |
Year	Hybrid	I	II	III	IV	I	II	III	IV
1992	Pioneer 3737	135.4	132.0	133.9	136.1	125.2	122.4	126.0	122.8
	Pioneer 3744	100.9	101.8	102.0	105.3	110.0	109.6	106.4	108.5
	Funk G-4312	105.0	104.9	102.1	106.3	111.3	110.2	110.9	111.0
	Funk G-4342	122.5	121.3	122.9	122.0	110.9	110.2	111.2	110.5
	DeKalb 484	128.2	127.6	125.9	126.3	123.9	124.0	124.1	124.2
	DeKalb 524	126.8	127.2	126.9	128.0	133.1	132.5	131.9	132.3
	Cargil 842	138.1	137.6	137.2	135.7	120.0	123.3	124.7	120.5
1993	Pioneer 3737	133.4	132.0	133.9	131.1	118.2	126.4	123.0	123.7
	Pioneer 3744	117.9	115.8	112.0	115.5	120.0	129.7	126.3	128.4
	Funk G-4312	109.0	108.9	105.3	108.7	108.3	106.4	109.9	107.0
	Funk G-4342	104.5	101.3	106.9	107.4	120.9	122.2	121.2	120.8
	DeKalb 484	124.8	126.6	125.3	124.9	113.8	114.0	114.1	114.2
	DeKalb 524	136.9	137.2	136.9	140.0	132.1	136.5	131.3	135.8
	Cargil 842	108.9	107.6	107.2	105.8	118.3	119.3	120.7	121.4

(a) Perform a combined analysis over the years.
(b) Perform a test of homogeneity of variance. Can the hypothesis of homogeneous
variances be accepted?
(c) What can be said about the nature of the interaction effect between the hybrid
and tillage interaction?

Solution:

a.

To perform the combined analysis, we first need to perform a separate analysis of
variance for each year in the experiment. This is done for the years 1992, and 1993.

Replication x tillage method (Factor *A*) Table of Totals

Replication	Conventional tillage (CT)	No tillage(NT)	Rep. Total (*R*)
I	856.9	834.4	1691.3
II	852.4	832.2	1684.6
III	850.9	835.2	1686.1
IV	859.7	829.8	1689.5
Tillage Total (*A*)	3419.9	3331.6	
Grand Total			6751.5

Hybrid x Tillage system table of Totals

Hybrid	CT	NT	hybrid Total (B)
Pioneer 3737	537.4	496.4	1033.8
Pioneer 3744	410.0	434.5	844.5
Funk G-4312	418.3	443.4	861.7
Funk G-4342	488.7	442.8	931.5
DeKalb 484	508.0	496.2	1004.2
DeKalb 524	508.9	529.8	1038.7
Cargil 842	548.6	488.5	1037.1

The following sources of variation and the degrees of freedom are identified.

$$\text{Replication } d.f. \quad = r - 1 \quad\quad = 3$$

$$\text{Treatment } d.f. \quad = ab - 1 \quad\quad = 14$$

$$\text{Tillage } (A) \, d.f. \quad = a - 1 \quad\quad = 1$$

$$\text{Hybrid } (B) \, d.f. \quad = b - 1 \quad\quad = 6$$

$$A \times B \, d.f. \quad = (a - 1)(b - 1) \quad = 6$$

$$\text{Error } d.f. \quad = (r - 1)(ab - 1) = 39$$

$$\text{Total } d.f. \quad = (rab - 1) \quad\quad = 55$$

The correction factor is:

$$C = \frac{G^2}{rab}$$

$$= \frac{(6751.50)^2}{4 \times 2 \times 7} = 813977.72$$

$$\text{Total} SS = \sum X^2 - C$$

$$= [(135.4)^2 + (132)^2 + \ldots + (120.5)^2] - 813977.72$$

$$= 6626.09$$

$$\text{Replication} SS = \frac{\sum R^2}{ab} - C$$

$$= \frac{(1691.3)^2 + (1684.6)^2 + (1686.1)^2 + (1689.5)^2}{14} - 813977.72$$

$$= 2.02$$

$$SS_A \text{ (Tillage)} = \frac{\sum A^2}{rb} - C$$

$$= \frac{(3419.9)^2 + (3331.6)^2}{4 \times 7} - 813977.72$$

$$= 139.23$$

$$\text{Error } SS\,(a) = \frac{\sum (RA)^2}{b} - (C + \text{Rep. } SS + SS_A)$$

$$= \frac{(856.9)^2 + (834.4)^2 + \ldots + (829.8)^2}{7} - (813977.72 + 2.02 + 139.23)$$

$$= 7.54$$

$$SS_B \text{ (Hybrid)} = \frac{\sum B^2}{ra} - C$$

$$= \frac{(1033.8)^2 + (844.5)^2 + \ldots + (1037.10)^2}{4 \times 2} - 813977.72$$

$$= 5401.48$$

$$SS_{AB} \text{ (Tillage x Hybrid)} = \frac{\sum (AB)^2}{r} - (C + SS_A + SS_B)$$

$$= \frac{(537.4)^2 + (496.4)^2 + \ldots + (488.5)^2}{4} - (813977.72 + 139.23 + 5401.48)$$

$$= 1011.54$$

$$\text{Error } SS\,(b) = TSS - \text{All other sum of squares}$$

$$= 6626.09 - (2.02 + 139.23 + 7.54 + 5401.48 + 1011.54)$$

$$= 64.28$$

The ANOVA table shown below presents the analysis for the years 1992 and 1993. Computations were only carried out for 1992.

<u>ANOVA Table for separate years</u>

Source of Variation	Degree of Freedom	Sum of Squares	Mean Square	F
1992				
Replication	3	2.02	0.67	
Hybrid (A)	6	139.23	23.21	9.23*
Error (a)	3	7.54	2.51	
Tillage (B)	1	5401.48	5401.48	3024.66**
Hybrid x tillage	6	1011.54	168.59	94.40**
Error (b)	36	64.29	1.79	
Total	55	6626.09		
1993				
Replication	3	15.37	5.12	0.44^{ns}
Hybrid (A)	6	60.49	10.08	0.88^{ns}
Error (a)	3	34.22	11.41	
Tillage (B)	1	4347.11	4347.11	1078.42**
Hybrid x tillage	6	1495.08	249.18	61.81**
Error (b)	36	145.12	4.03	
Total	55	6097.39		

ns= nonsignificant, *= Significant at the 5% level, ** = Significant at the 1% level.

Determine the degrees of freedom associated with each source of variation. Since in the present case year (Y) is considered a random variable, the error term is represented by the year x irrigation interaction MS. Thus, we have:

$$\text{Year } (Y)\ d.f. = y - 1 = 2$$

$$\text{Reps. within year} = y(r - 1) = 9$$

$$\text{Irrigation} = t - 1 = 3$$

$$\text{Year x Irrigation} = (y - 1)(t - 1) = 6$$

$$\text{Pooled error} = y(r - 1)(t - 1) = 27$$

$$\text{Total } d.f. = yrt - 1 = 47$$

Compute the sum of squares for the replications within the year and the pooled error as follows:

$$\text{Rep. within year } SS = \sum_{i=1}^{y} (\text{Rep. } SS)_i$$

$$\text{Pooled Error } SS = \sum_{i=1}^{y} (\text{Error } SS)_i$$

Thus, we have:

$$\text{Reps. within year } SS = 15.37 + 2.02 = 17.39$$

$$\text{Pooled error } SS = 7.54 + 34.22 = 41.76$$

Calculate the correction factor as:

$$C = \frac{\left(\sum\limits_{i=1}^{y} G_i\right)^2}{yrab}$$

$$C = \frac{(6751.50 + 6709.60)^2}{112} = 1617867.98$$

Compute the sum of squares for the year, treatment, and the year x treatment interaction as follows:

$$\text{Year } SS = \sum_{i=1}^{y} \frac{G_i^2}{rab} - C$$

$$\text{Year } SS = \frac{(6751.50)^2 + (6709.60)^2}{4(14)} - 1617867.98$$
$$= 15.68$$

$$\text{Treatment } SS = \sum_{j=1}^{t} \frac{T_j^2}{yr} - C$$

$$= \frac{\left[(2055.5)^2 + (1810.10)^2 + \ldots + (1946.3)^2\right]}{4(2)} - 1617867.98$$

$$= 1632855.10$$

$$\text{Year x Treatment } SS = \sum_{i=1}^{y} \sum_{j=1}^{t} \frac{(YT)_{ij}^2}{r} - (C + \text{Year } SS - \text{Treatment } SS)$$

Where:

$$(YT)_{ij} = \text{the total of the } j \text{ th treatment in the } i \text{ th year.}$$

$$\text{Year x Treatment } SS = \frac{\left[(1033.8)^2 + (844.5)^2 + \ldots + (909.20)^2\right]}{4} - (1617867.98 + 15.68 + 7493.57)$$

$$= 2255.02$$

Combined ANOVA Table for a RCB Experiment over 2 years

Source of Variation	Degree of Freedom	Sum of Squares	Mean Square	F
Year	1	15.68	15.68	13.51**
Reps. within yr.	6	17.39	2.89	2.49*
Treatment	6	7493.57	1248.93	1076.71**
Year x Treatment	6	2255.03	375.84	324.01**
Pooled Error	36	41.76	1.16	
Total	55	9823.42		

8= Significant at the 5% level, **= Significant at the 1% level.

The results show that all effects and interaction are significant.

b. To perform a test of homogeneity of variance, we apply the following F test. We could have just as well used the chi-square test.

$$F = \frac{\text{Larger error } MS}{\text{Smaller error } MS}$$

$$F = \frac{4.03}{1.79} = 2.25$$

In comparing the computed F value with the corresponding F value (1.76 at the 5% level) found in Appendix E for 36 degrees of freedom, we reject the hypothesis of homogeneous error variances. Since we rejected the hypothesis of homogeneous error variance, the pooled error variance should be partitioned for further analysis.

c. As seen in the ANOVA table, the interaction effect is highly significant.

3. In an attempt to determine the effect of harvest management on forage yield from "Nitro" a cultivar of alfalfa (*Medicago sativa* L.), a randomized complete block experiment with 4 replicates was conducted. The treatments consisted of the following 4 harvest management systems:

H_1: No harvest during the growing season;
H_2: Two harvests at bud and herbage regrowth harvested in the fall;
H_3: Three harvests at bud and herbage regrowth harvested in the fall;
H_4: Two harvests at first flower and herbage regrowth harvested in the fall

The data on forage yield collected for the Summer and Fall are given below.

Effect of harvest management on forage yield (ton/acre)

Harvest Management	Rep. I	Rep. II	Rep. III	Rep. IV	Total
			Summer		
H_1	--	--	--	--	0.0
H_2	1.9	2.0	2.1	2.0	8.0
H_3	3.0	3.1	3.0	2.8	11.9
H_4	2.5	2.4	2.8	2.3	10.0
Total	7.4	7.5	7.9	7.1	29.9
			Fall		
H_1	0.5	2.3	1.4	2.1	6.3
H_2	0.7	0.9	0.8	0.7	3.1
H_3	0.4	0.6	0.2	0.1	1.3
H_4	0.3	0.5	0.3	0.4	1.5
Total	1.9	4.3	2.7	3.3	12.2

(a) Perform a combined analysis over the seasons.

(b) Perform a test of homogeneity of variance. Can the hypothesis of homogeneous variances be accepted?

(c) What can be said about the nature of interaction effect between the seasons and harvest management practices?

Solution:

a. For simplicity, we have added the total for the replications and total for different harvest management in the table above.

Results of the Analyses of Variance for the Individual Seasons Using an RCB Design with 4 Harvest Management Schemes and 4 Replications

Source of Variation	Degree of Freedom	Sum of Squares	Mean Square	F
		Summer		
Replication	3	0.082	0.027	
Harvest	3	20.527	6.842	488.71**
Error	9	0.126	0.014	
		Fall		
Replication	3	0.768	0.256	
Harvest	3	4.008	1.336	8.45**
Error	9	1.423	0.158	

** = significant at 1% level.

Determine the degrees of freedom associated with each source of variation based on the RCB design used in the experiment. The following sources of variation are identified:

Season (S) $d.f.$	$= s - 1$	$=$	1
Reps. within season	$= s(r - 1)$	$=$	6
Treatment (Harvest)	$= t - 1$	$=$	3
Season x Harvest	$= (s - 1)(t - 1)$	$=$	3
Pooled error	$= s(r - 1)(t - 1)$	$=$	18
Total $d.f.$	$= srt - 1$	$=$	31

Compute the sum of squares for the replications within the season and the pooled error as follows:

Rep. within season $SS = \sum_{i=1}^{s} (\text{Rep. } SS)_i$

Pooled Error $SS = \sum_{i=1}^{s} (\text{Error } SS)_i$

Thus we have:

Reps. within season $SS = 0.082 + 0.768 = 0.850$

Pooled error $SS = 0.126 + 1.423 = 1.549$

The correction factor is:

$$C = \frac{\left(\sum_{i=1}^{s} G_i\right)^2}{srt}$$

$$C = \frac{(29.9 + 12.20)^2}{32} = 55.39$$

$$\text{Season } SS = \sum_{i=1}^{s} \frac{G_i^2}{rt} - C$$

$$\text{Season } SS = \frac{(29.9)^2 + (12.2)^2}{4(4)} - 55.39$$

$$= 9.79$$

$$\text{Harvest } SS = \sum_{j=1}^{t} \frac{T_j^2}{sr} - C$$

$$= \frac{\left[(6.3)^2 + (11.1)^2 + (13.2)^2 (11.5)^2\right]}{2(4)} - 55.39$$

$$= 3.29$$

$$\text{Season x Harvest } SS = \sum_{i=1}^{s} \sum_{j=1}^{t} \frac{(st)_{ij}^{2}}{r} - (C + \text{Season } SS + \text{Harvest } SS)$$

Where:

$$(st)_{ij} = \text{the total of the } j \text{ th treatment in the } i \text{ th season.}$$

$$\text{Season x Harvest } SS = \frac{\left[(0.0)^2 + (64)^2 + \ldots + (2.25)^2\right]}{4} - (55.39 + 9.79 + 3.29)$$

$$= 21.25$$

Combined Analysis of Variance of a Randomized Complete Block Experiment over 2 Planting Seasons

Source of Variation	Degree of Freedom	Sum of Squares	Mean Square	F
Season	1	9.79	9.79	122.37**
Reps. within season	6	0.85	0.14	1.75*ns*
Harvest	3	3.29	1.09	13.63**
Season x Harvest	3	21.25	7.08	88.50**
Pooled Error	18	1.55	0.08	
Total	31	36.73		

ns=Nonsignificant, **= highly significant at the 1% level.

b. To test the homogeneity of variance we use the following:

$$F = \frac{\text{Larger error } MS}{\text{Smaller error } MS}$$

$$F = \frac{0.158}{0.014} = 11.29$$

In comparing the computed F value with the corresponding F value (5.35 at the 1% level) found in Appendix E for 9 degrees of freedom, we reject the hypothesis of homogeneous error variances over seasons. Thus, we need to partition Season x Harvest interaction into a set of orthogonal contrasts.

Since there are 4 treatments, the contrasts from Appendix J are:

Harvest Management	Orthogonal Polynomial Coefficient		
	Linear	Quadratic	Cubic
H1	-3	1	-1
H2	-1	-1	3
H3	1	-1	-3
H4	3	1	1

To compute the sum of squares for each of these single degree of freedom contrast we need to construct a Harvest x Season Table of Totals

Harvest	Rep. I	Rep. II	Rep. III	Rep. IV	Harvest Total
H1	0.5	2.3	1.4	2.1	6.3
H2	2.6	2.9	2.9	2.7	11.1
H3	3.4	3.7	3.2	2.9	13.2
H4	2.8	2.9	3.1	2.7	11.5

$$\text{Harvest}_{\text{Linear}}\ SS = \frac{\left[(-3)(H_1)+(-1)(H_2)+(1)(H_3)+(3)(H_4)\right]^2}{(r)(a)\left[(-3)^2+(-1)^2+(1)^2+(3)^2\right]}$$

$$\text{Harvest}_{\text{Quadratic}}\ SS = \frac{\left[(1)(H_1)+(-1)(H_2)+(-1)(H_3)+(1)(H_4)\right]^2}{(r)(a)\left[(1)^2+(-1)^2+(-1)^2+(1)^2\right]}$$

$$\text{Harvest}_{\text{Cubic}}\ SS = \frac{\left[(-1)(H_1)+(3)(H_2)+(-3)(H_3)+(1)(H_4)\right]^2}{(r)(a)\left[(-1)^2+(3)^2+(-3)^2+(1)^2\right]}$$

Where

$H_1 \ldots H_4$ = Harvest totals

r = number of replications

a = number of seasons.

$$\text{Harvest}_{\text{Linear}}\ SS = \frac{\left[(-3)(6.3)+(-1)(11.1)+(1)(13.2)+(3)(11.5)\right]^2}{(2)(4)\left[(-3)^2+(-1)^2+(1)^2+(3)^2\right]} = 1.958$$

$$\text{Harvest}_{\text{Quadratic}}\ SS = \frac{\left[(1)(6.3)+(-1)(11.1)+(-1)(13.2)+(1)(11.5)\right]^2}{(4)(2)\left[(1)^2+(-1)^2+(-1)^2+(1)^2\right]} = 1.32$$

$$\text{Harvest}_{\text{Cubic}}\ SS = \frac{\left[(-1)(6.3)+(3)(11.1)+(-3)(13.2)+(1)(11.5)\right]^2}{(4)(2)\left[(-1)^2+(3)^2+(-3)^2+(1)^2\right]} = 0.008$$

Harvest Total and the Linear, Quadratic and Cubic Sum of Squares Computed from Data

Season	Harvest Total				Sum of Squares		
	H_1	H_2	H_3	H_4	Linear	Quadratic	Cubic
Summer	0.0	8.0	11.9	10.0	14.37	6.13	0.04
Fall	6.3	3.1	1.3	1.5	3.28	0.72	0.01

Compute the sum of squares for each contrast based on the harvest totals at each season. The computation is shown below for the summer only, however, the same procedure is applied in computing the data for fall. The results are shown in the above table for both seasons.

$$\text{Harvest}_{\text{Linear}}\ SS = \frac{\left[(-3)(0)+(-1)(8)+(1)(11.9)+(3)(10)\right]^2}{(4)\left[(-3)^2+(-1)^2+(1)^2+(3)^2\right]} = 14.37$$

$$\text{Harvest}_{\text{Quadratic}}\ SS = \frac{\left[(1)(0)+(-1)(8)+(-1)(11.9)+(1)(10)\right]^2}{(4)\left[(1)^2+(-1)^2+(-1)^2+(1)^2\right]} = 6.13$$

$$\text{Harvest}_{\text{Cubic}}\ SS = \frac{\left[(-1)(0)+(3)(8)+(-3)(11.9)+(1)(10)\right]^2}{(4)\left[(-1)^2+(3)^2+(-3)^2+(1)^2\right]} = 0.04$$

Calculate the component of the Harvest x Season interaction corresponding to the set of mutually orthogonal contrasts as follows:

$$\text{Harvest}_i \ x\ \text{Season } SS = \sum_{j=1}^{s} (\text{Harvest}_i\ SS)_j - \text{Harvest}_i\ SS$$

Harvest Linear x Season SS = (14.37+3.281) -1.958 =15.693

Harvest Quadratic x Season SS = (6.125+0.723) -1.320 = 5.528

Harvest Cubic x Season SS = (0.0361+0.0045) - 0.0076 = 0.033

Finally the results are shown in the ANOVA Table below.

Analysis of Variance of a Partitioned Treatment Sum of Squares of a Randomized Complete Block Experiment

Source of Variation	Degree of Freedom	Sum of Squares	Mean Square	F
Season	1	9.79	9.77	9.88
Reps. within season	6	0.85	0.14	1.65
Harvest	(3)	(3.29)	1.09	12.74**
$\quad$ Harvest$_L$	1	1.96	1.96	22.77**
$\quad$ Harvest$_Q$	1	1.32	1.32	15.35**
$\quad$ Harvest$_C$	1	0.01	0.01	0.09^{ns}
Harvest x Season	(3)	(21.25)	7.08	82.35**
$\quad$ Harvest$_L$ x Season	1	15.69	15.69	182.40**
$\quad$ Harvest$_Q$ x Season	1	5.53	5.53	64.26**
$\quad$ Harvest$_C$ x Season	1	0.03	0.03	0.38^{ns}
Pooled error	18	1.55	0.09	
Total	31			

^{ns} = nonsignificant, ** = significant at 1% level.

The analysis shows that the linear component of the sum of squares varied significantly with season. This implies that the Season x Harvest interaction is mainly due to the difference in the linear part of the yield responses to harvest management of the different seasons. We also note that both the linear and the quadratic component of the Season x Harvest interaction are significant. This implies that forage yield initially increased at lower rates, tapered off at a maximum, and finally decreased at higher harvest rates.

The results indicate that different harvest management technique need to be used for the summer and fall seasons. The average forage yield response to different harvesting strategies between seasons is important in making accurate economic harvesting strategy recommendations that are environmentally sound.

Since the test of homogeneity is significant, then the appropriate test of significance requires that the error term be the component of the pooled error and not the pooled error itself.

Partition the pooled error sum of squares into components corresponding to those of the Season x Harvest sum of squares as:

$$\text{Reps. within Seas. x Harvest}_{Lin} = \sum_{i=1}^{s} \left[\sum_{j=1}^{r} (\text{Harvest}_{Lin}SS)_{ji} - (\text{Harvest}_{Lin}SS)_{i} \right]$$

$$\text{Reps. within Seas. x Harvest}_{Quad} = \sum_{i=1}^{s} \left[\sum_{j=1}^{r} (\text{Harvest}_{Quad}SS)_{ji} - (\text{Harvest}_{Quad}\,SS)_{i} \right]$$

$$\text{Reps. within Seas. x Harvest}_{Cubic} = \sum_{i=1}^{s} \left[\sum_{j=1}^{r} (\text{Harvest}_{Cubic}SS)_{ji} - (\text{Harvest}_{Cubic}\,SS)_{i} \right]$$

We need to compute the (Harvest linear SS)$_{ij}$ as shown below.

$$\text{Harvest}_{Lin}\,SS = \frac{\left[(-3)(H_1)+(-1)(H_2)+(1)(H_3)+(3)(H_4)\right]^2}{\left[(-3)^2+(-1)^2+(1)^2+(3)^2\right]}$$

$$\text{Harvest}_{Lin}\,SS = \frac{\left[(-3)(0)+(-1)(1.9)+(1)(3)+(3)(2.5)\right]^2}{\left[(-3)^2+(-1)^2+(1)^2+(3)^2\right]} = 3.698$$

$$\text{Harvest}_{Quad}\,SS = \frac{\left[(1)(H_1)+(-1)(H_2)+(-1)(H_3)+(1)(H_4)\right]^2}{\left[(1)^2+(-1)^2+(-1)^2+(1)^2\right]}$$

$$\text{Harvest}_{Quad}\,SS = \frac{\left[(1)(0)+(-1)(1.9)+(-1)(3)+(1)(2.5)\right]^2}{\left[(1)^2+(-1)^2+(-1)^2+(1)^2\right]} = 1.44$$

$$\text{Harvest}_{Cubic}\,SS = \frac{\left[(-1)(H_1)+(3)(H_2)+(-3)(H_3)+(1)(H_4)\right]^2}{\left[(-1)^2+(3)^2+(-3)^2+(1)^2\right]}$$

$$\text{Harvest}_{Cubic}\,SS = \frac{\left[(-1)(0)+(3)(1.9)+(-3)(3)+(1)(2.5)\right]^2}{\left[(-1)^2+(3)^2+(-3)^2+(1)^2\right]} = 0.032$$

We have used the above equations to compute the values for all replications. The results are shown in the table below.

Treatment Totals and Linear, Quadratic and Cubic Sum of Squares for the Pooled Error Sum of Squares

| Replication | Harvest Total | | | | Sum of Squares | | |
	H_1	H_2	H_3	H_4	Linear	Quadratic	Cubic
		Summer					
I	0.0	1.9	3.0	2.5	3.698	1.440	0.0320
II	0.0	2.0	3.1	2.4	3.445	1.823	0.0405
III	0.0	2.1	3.0	2.8	4.325	1.323	0.0005
IV	0.0	2.0	2.8	2.3	2.965	1.563	0.0005
Total	0.0	8.0	11.9	10.0	14.370	6.130	0.0360
		Fall					
I	0.5	0.7	0.4	0.3	0.041	0.023	0.0245
II	2.3	0.9	0.6	0.5	1.625	0.423	0.0405
III	1.4	0.8	0.2	0.3	0.761	0.123	0.0245
IV	2.1	0.7	0.1	0.4	1.625	0.723	0.0005
Total	6.3	3.1	1.3	1.5	3.281	0.720	0.0050

Now we are ready to compute the components of the pooled error sum of squares as:

$$\text{Reps. within Seas. x Harvest}_{Lin} = \left[(3.69 + 3.45 + 4.33 + 2.97) - 14.37 \right] + \left[(0.04 + 1.63 + 0.76 + 1.63) - 3.28 \right]$$

$$= 0.836$$

$$\text{Reps. within Seas. x Harvest}_{Quad} = \left[(1.44 + 1.82 + 1.32 + 1.56) - 6.13 \right] + \left[(0.02 + 0.42 + 0.12 + 0.72) - 0.72 \right]$$

$$= 0.59$$

$$\text{Reps. within Seas. x Harvest}_{Cubic} = \left[(0.032 + 0.041 + 0.001 + 0.001) - 0.036 \right] + \left[(0.024 + 0.041 + 0.024 + 0.001) - 0.005 \right]$$

$$= 0.123$$

The results are summarized in the table below.

Component of the Pooled Error Sum of Squares

Source of Variation	D.F.	SS	MS
Pooled Error	(18)	(1.548)	0.086
Reps. within Season x Harvest $_{Lin}$	6	0.836	0.139
Reps. within Season x Harvest $_{Quad}$	6	0.590	0.098
Reps. within Season x Harvest $_{Cubic}$	6	0.123	0.020

Test the significance of the Season x Harvest interaction in the following way.

$$F_{Harvest_{Lin} \text{ x Season}} = \frac{Harvest_{Lin} \text{ x Season} MS}{\text{Reps. within Season x Harvest}_{Lin} MS}$$

$$F_{Harvest_{Quad} \text{ x Season}} = \frac{Harvest_{Quad} \text{ x Season} MS}{\text{Reps. within Season x Harvest}_{Quad} MS}$$

$$F_{Harvest_{Cubic} \text{ x Season}} = \frac{Harvest_{Cubic} \text{ x Season} MS}{\text{Reps. within Season x Harvest}_{Cubic} MS}$$

$$F_{Harvest_{Lin} \text{ x Season}} = \frac{15.69}{0.139} = 112.88$$

$$F_{Harvest_{Quad} \text{ x Season}} = \frac{5.53}{0.098} = 56.43$$

$$F_{Harvest_{cubic} \text{ x Season}} = \frac{0.033}{0.02} = 1.65$$

Again the linear and quadratic component are highly significant. The interaction effects are highly significant.

4. Since grazing of hay fields during some part of the year is used as an alternative to harvested hay for beef cattle, an animal scientist compared 4 management systems to determine the yield of forage produced during the spring and fall. The data gathered from the randomized complete block experiment with 5 replications are shown below.

Total seasonal herbage dry matter harvested as influenced by management (lb/acre)

Management	Rep. I	Rep. II	Rep. III	Rep. IV	Rep. V
			Spring		
Hay only	4015	4020	4000	3995	4005
Spring and fall grazing	3756	3776	3789	3886	3554
Spring grazing	3300	3312	3310	3345	3305
Fall grazing	4583	4539	4601	4520	4495
			Fall		
Hay only	2334	2486	2300	2319	2398
Spring and fall grazing	3300	3198	3264	3310	3315
Spring grazing	2404	2400	2445	2487	2404
Fall grazing	3360	3354	3329	3352	3390

(a)	Perform a combined analysis over the seasons.
(b)	Perform a test of homogeneity of variance. Can the hypothesis of homogeneous variances be accepted?
(c)	What can be said about the nature of the interaction effect between the seasons and the management systems?

Solution:

a.

Solution:

To avoid large numbers in the computation, the original have been divided 1000. The results are given below.

Total seasonal herbage dry matter harvested as influenced by management (lb/acre)

Management	Rep. I	Rep. II	Rep. III	Rep. IV	Rep. V	Total
			Spring			
Hay only	4.015	4.020	4.000	3.995	4.005	20.035
Spring and fall grazing	3.756	3.776	3.789	3.886	3.554	18.761
Spring grazing	3.300	3.312	3.310	3.345	3.305	16.572
Fall grazing	4.583	4.539	4.601	4.520	4.495	22.738
Total	15.650	15.650	15.700	15.750	15.360	78.106
			Fall			
Hay only	2.334	2.486	2.300	2.319	2.398	11.837
Spring and fall grazing	3.300	3.198	3.264	3.310	3.315	16.387
Spring grazing	2.404	2.400	2.445	2.487	2.404	12.14
Fall grazing	3.360	3.354	3.329	3.352	3.390	16.785
Total	11.400	11.440	11.340	11.470	11.510	57.149

The procedure for performing the individual analysis of variance is similar to that of Exercise 8-2 and need not be repeated. The results of the analysis are presented in the table below.

Results of the Analyses of Variance for the Individual Seasons Using an RCB Design with 4 Harvest Management Schemes and 4 Replications

Source of Variation	Degree of Freedom	Sum of Squares	Mean Square	F
		Spring		
Replication	4	0.023	0.0058	1.52
Management	3	3.977	1.3258	350.08**
Error	12	0.045	0.0038	
		Fall		
Replication	4	0.004	0.0011	0.35
Management	3	4.252	1.4175	474.27**
Error	12	0.036	0.003	

** = significant at 1% level.

Determine the degrees of freedom associated with each source of variation based on the RCB design used in the experiment. The following sources of variation are identified:

Season (S) d.f. $\quad = \quad s - 1 \quad = \quad 1$
Reps. within season $\quad = \quad s\,(r - 1) \quad = \quad 8$
Treatment (Management) $\quad = \quad t - 1 \quad = \quad 3$
Season x Management $\quad = \quad (s - 1)(t - 1) \quad = \quad 3$
Pooled error $\quad = \quad s\,(r - 1)(t - 1) \quad = \quad 24$
Total d.f. $\quad = \quad srt - 1 \quad = \quad 39$

Compute the sum of squares for the replications within the season and the pooled error as follows:

$$\text{Rep. within season } SS = \sum_{i=1}^{s} (\text{Rep. } SS)_i$$

$$\text{Pooled Error } SS = \sum_{i=1}^{s} (\text{Error } SS)_i$$

Thus we have:

Reps. within season $SS = 0.023 + 0.004 = 0.027$

Pooled error $SS = 0.045 + 0.036 = 0.081$

The correction factor is:

$$C = \frac{\left(\sum\limits_{i=1}^{s} G_i\right)^2}{srt}$$

$$C = \frac{(78.106 + 57.149)^2}{40} = 457.30$$

$$\text{Season } SS = \sum_{i=1}^{s} \frac{G_i^{\,2}}{rt} - C$$

$$\text{Season } SS = \frac{(78.106)^2 + (57.15)^2}{5(4)} - 457.30$$

$$= 10.98$$

$$\text{Management } SS = \sum_{j=1}^{t} \frac{T_j^{\,2}}{sr} - C$$

$$= \frac{\left[(31.87)^2 + (35.15)^2 + (28.71)^2 (39.52)^2\right]}{2(5)} - 457.30$$

$$= 6.42$$

$$\text{Season x Management } SS = \sum_{i=1}^{s} \sum_{j=1}^{t} \frac{(st)_{ij}^2}{r} - (C + \text{Season } SS + \text{Management } SS)$$

Where:

$$(st)_{ij} = \text{the total of the } j \text{ th treatment in the } i \text{ th season.}$$

$$\text{Season x Management } SS = \frac{\left[(401.4)^2 + (351.98)^2 + \ldots + (281.74)^2\right]}{5} - (457.30 + 10.98 + 6.42)$$

$$= 1.81$$

Combined Analysis of Variance of a Randomized Complete Block Experiment over 2 Planting Seasons

Source of Variation	Degree of Freedom	Sum of Squares	Mean Square	F
Season	1	10.980	10.980	
Replication within Sea	8	0.027	0.003	1.007ns
Management	3	6.417	2.139	631.39**
Season x Mgmt.	3	1.812	0.604	178.33**
Pooled Error	24	0.081	0.003	
Total	39			

ns=Nonsignificant, **= highly significant at the 1% level.

b. To test the homogeneity of variance we use the following:

$$F = \frac{\text{Larger error } MS}{\text{Smaller error } MS}$$

$$F = \frac{0.004}{0.003} = 1.33$$

As the computed value of 1.33 is less than the table of value of 2.69 for the 12 degrees of freedom, we accept the hypothesis of homogeneous error variance. Thus, we do not need to partition season x management interactions.

c. As we can see, both the management main effect and its interaction with crop season are highly significant; there is a significant yield response to how fields are managed (different grazing pattern by animals).

Chapter 9

REGRESSION AND CORRELATION ANALYSIS

1. A researcher interested in the relationship between the rate of germination
of warm-season forage grasses and temperature has postulated a linear regression model
where the number of seeds germinated per day is dependent on the average daily
temperature. The following data were collected by the researcher.

Germinated Seed (no./day)	Temperature (°C)	Germinated Seed (no./day)	Temperature (°C)
5	10	28	22
7	11	31	23
9	13	35	24
10	15	38	26
14	16	49	27
20	18	55	29
24	20	61	30
25	21	73	32

(a) Use the least-square technique in estimating the equation.
(b) Compute the standard error of estimate. What is the interpretation of the standard error of estimate?
(c) Is there a significant correlation between the 2 variables? Use a 5% level of significance.

Solution:

(a). The least-square technique requires the computation of the intermediate values as shown below:

Germinated Seed (no./day) Y	Temperature (°C) X	Y^2	XY	X^2
5	10	25	50	100
7	11	49	77	121
9	13	81	117	169
10	15	100	150	225
14	16	196	224	256
20	18	400	360	324
24	20	576	480	400
25	21	625	525	441
28	22	784	616	484
31	23	961	713	529
35	24	1225	840	576
38	26	1444	988	676
49	27	2401	1323	729
55	29	3025	1595	841
61	30	3721	1830	900
73	32	5329	2336	1024
484	337	20942	12224	7795

The normal equation to be solved simultaneously are:

$$484 = 16a + 337b$$

$$12224 = 337a + 7795b$$

We multiply the first equation with 21.06 to get the following

$$10194.25 = 337a + 7098.06b \qquad (1)$$
$$-\ 12224 = 337a + 7795b \qquad\qquad (2)$$

Subtract equation 2 from equation 1. variable a cancel each other out, and we get:

-2029.75 = - 696.94b

$b= 2.91$

Now we can substitute the value of b into the first equation and solve for a:

$$484 = 16a + 337(2.91)$$

$$a = -31.04$$

Our estimated equation is:

$$\hat{Y} = -31.04 + 2.91X$$

b. To compute the standard error of estimate we first determine the intermediate values as shown in the table below:

Y	X	$\hat{Y}$	$Y - \hat{Y}$	$(Y - \hat{Y})^2$
5	10	-1.94	6.94	48.164
7	11	0.97	6.03	36.361
9	13	6.79	2.21	4.884
10	15	12.61	-2.61	6.812
14	16	15.52	-1.52	2.310
20	18	21.34	-1.34	1.796
24	20	27.16	-3.16	9.986
25	21	30.07	-5.07	25.705
28	22	32.98	-4.98	24.800
31	23	35.89	-4.89	23.912
35	24	38.8	-3.80	14.440
38	26	44.62	-6.62	43.824
49	27	47.53	1.47	2.161
55	29	53.35	1.65	2.723
61	30	56.26	4.74	22.468
73	32	62.08	10.92	119.250
484	337	484.00	0.0	389.590

$$S_{yx} = \sqrt{\frac{\Sigma(Y - \hat{Y})^2}{n-2}}$$

$$S_{yx} = \sqrt{\frac{389.6}{14}}$$

$$S_{yx} = 5.27$$

Since the standard error of estimate is theoretically similar to the standard deviation, there is also similarity in the interpretation of the standard error of estimate. If the scatter about the regression line is normally distributed, and we have a large sample, approximately 68% of the points in the scatter diagram will fall within 1 standard error of estimate. We use the standard error of estimate in estimating the confidence interval.

c. To determine if there is correlation between the seed germination and the temperature, we compute the coefficient of determination first, and then use this value to determine the correlation coefficient. The mean for this problem is 30.25

Y	X	$\hat{Y}$	$(Y-\bar{Y})^2$	$(\hat{Y}-\bar{Y})^2$
5	10	-1.94	637.56	1036.20
7	11	0.97	540.56	857.32
9	13	6.79	451.56	550.37
10	15	12.61	410.06	311.17
14	16	15.52	264.06	216.97
20	18	21.34	105.06	79.39
24	20	27.16	39.06	9.55
25	21	30.07	27.56	0.03
28	22	32.98	5.06	7.45
31	23	35.89	0.56	31.81
35	24	38.8	22.56	73.10
38	26	44.62	60.06	206.50
49	27	47.53	351.56	298.60
55	29	53.35	612.56	533.61
61	30	56.26	945.56	676.52
73	32	62.08	1827.60	1013.10
484	337	484.00	6301.00	5901.70

c. The coefficient of determination is computed as:

$$r^2 = \frac{5901.7}{6301} = 0.936$$

$$r = \sqrt{0.936}$$

$$r = 0.967$$

The results show that there is a high degree of correlation between the two variable.

2. In a 2-factor experiment, an environmental horticulturist wishes to determine the impact of nitrogen fertilization and hourly exposure to sunlight on leaf thickness in the rubberplant. After 2 months of experimentation, the following average observations of leaf thickness, amount of light in the greenhouse, and fertilization were recorded from an experiment with 4 replications. Plants of the same age were grown in the same environment varying only the 2 factors of interest.

Treatment Number	Leaf thickness (mm)	Hours of sunlight	Nitrogen fertilization mg/pot
1	2.15	5.0	.20
2	2.28	5.0	.30
3	4.56	5.0	.40
4	7.68	5.0	.50
5	8.76	5.0	.60
6	12.80	5.0	.70
7	15.62	5.0	.80
8	16.54	5.0	.90
9	3.24	6.0	.20
10	6.32	6.0	.30
11	8.76	6.0	.40
12	7.48	6.0	.50
13	12.69	6.0	.60
14	16.87	6.0	.70
15	20.92	6.0	.80
16	18.66	6.0	.90
17	5.33	7.0	.20
18	6.38	7.0	.30
19	6.87	7.0	.40
20	8.88	7.0	.50
21	9.89	7.0	.60
22	20.87	7.0	.70
23	21.92	7.0	.80
24	22.96	7.0	.90

(a) Estimate the regression equation.
(b) Compute the coefficient of multiple determination.
(c) Test the significance of R^2.
(d) Compute the various coefficients of correlation.

Solution:

a.

Leaf thickness (mm) Y	Hours of sunlight X_1	Nitrogen fertilization mg/pot X_2	YX_1	X_1^2	X_1X_2	YX_2	X_2^2	Y^2
2.15	5	0.2	10.75	25	1.0	0.43	0.04	4.62
2.28	5	0.3	11.40	25	1.5	0.68	0.09	5.20
4.56	5	0.4	22.80	25	2.0	1.82	0.16	20.79
7.68	5	0.5	38.40	25	2.5	3.84	0.25	58.98
8.76	5	0.6	43.80	25	3.0	5.26	0.36	76.74
12.8	5	0.7	64.00	25	3.5	8.96	0.49	163.84

15.62	5	0.8	78.10	25	4.0	12.49	0.64	243.98
16.54	5	0.9	82.70	25	4.5	14.89	0.81	273.57
3.24	6	0.2	19.44	36	1.2	0.65	0.04	10.49
6.32	6	0.3	37.92	36	1.8	1.90	0.09	39.94
8.76	6	0.4	52.56	36	2.4	3.50	0.16	76.74
7.48	6	0.5	44.88	36	3.0	3.74	0.25	55.95
12.69	6	0.6	76.14	36	3.6	7.61	0.36	161.04
16.87	6	0.7	101.22	36	4.2	11.81	0.49	284.59
20.92	6	0.8	125.52	36	4.8	16.74	0.64	437.65
18.66	6	0.9	111.96	36	5.4	16.79	0.81	348.19
5.33	7	0.2	37.31	49	1.4	1.07	0.04	28.41
6.38	7	0.3	44.66	49	2.1	1.91	0.09	40.70
6.87	7	0.4	48.09	49	2.8	2.75	0.16	47.19
8.88	7	0.5	62.16	49	3.5	4.44	0.25	78.85
9.89	7	0.6	69.23	49	4.2	5.93	0.36	97.81
20.87	7	0.7	146.09	49	4.9	14.61	0.49	435.56
21.92	7	0.8	153.44	49	5.6	17.54	0.64	480.49
22.96	7	0.9	160.72	49	6.3	20.66	0.81	527.16
268.4	144	13.2	1643.30	880	79.2	180.03	8.52	3998.50

$$\bar{Y} = 11.185 \qquad \bar{X}_1 = 6.00 \qquad \bar{X}_2 = 0.55$$

$$\Sigma y^2 = \Sigma Y^2 - \frac{(\Sigma Y)^2}{n} = 3998.5 - \frac{(268.4)^2}{24} = 996.2$$

$$\Sigma x_1^2 = \Sigma X_1^2 - \frac{(\Sigma X_1)^2}{n} = 880 - \frac{(144)^2}{24} = 16$$

$$\Sigma x_2^2 = \Sigma X_2^2 - \frac{(\Sigma X_2)^2}{n} = 8.52 - \frac{(13.2)^2}{24} = 1.26$$

$$\Sigma x_1 y = \Sigma X_1 Y - \frac{(\Sigma Y)(\Sigma X_1)}{n} = 1643.3 - \frac{(268.4)(144)}{24} = 32.71$$

$$\Sigma x_2 y = \Sigma X_2 Y - \frac{(\Sigma Y)(\Sigma X_2)}{n} = 180.03 - \frac{(268.4)(13.2)}{24} = 32.39$$

$$\Sigma x_1 x_2 = \Sigma X_1 X_2 - \frac{(\Sigma X_1)(\Sigma X_2)}{n} = 79.2 - \frac{(144)(13.2)}{24} = 0.0$$

$$b_1 = \frac{\left(\Sigma x_2^2\right)\left(\Sigma x_1 y\right) - \left(\Sigma x_1 x_2\right)\left(\Sigma x_2 y\right)}{\left(\Sigma x_1^2\right)\left(\Sigma x_2^2\right) - \left(\Sigma x_1 x_2\right)^2}$$

$$b_2 = \frac{\left(\Sigma x_1^2\right)\left(\Sigma x_2 y\right) - \left(\Sigma x_1 x_2\right)\left(\Sigma x_1 y\right)}{\left(\Sigma x_1^2\right)\left(\Sigma x_2^2\right) - \left(\Sigma x_1 x_2\right)^2}$$

$$b_1 = \frac{(1.26)(32.71) - (0)(32.3915)}{(16)(1.26) - (0)^2} = 2.04$$

$$b_2 = \frac{(16)(32.3915) - (0)(32.71)}{(16)(1.26) - (0)^2} = 25.71$$

$$a = \overline{Y} - b_1 \overline{X}_1 - b_2 \overline{X}_2$$

$$a = 11.185 - (2.04)(6.0) - (25.71)(0.55)$$

$$a = 15.19$$

(a). The estimated multiple linear regression is:

$$\hat{Y} = -15.19 + 2.04X_1 + 25.71X_2$$

(b). The coefficient of multiple determination is computed as:

$$R^2 = \frac{SSR}{SST}$$

Since SST is the same as Σy^2 which is 996.2, we need to compute the value for SSR before we can determine the coefficient of determination.

$$SSR = b_1 \Sigma x_1 y + b_2 \Sigma x_2 y$$

$$SSR = (2.04)(32.71) + (25.71)(32.3915) = 899.51$$

So the coefficient of determination is:

$$R^2 = \frac{899.51}{996.2} = 0.90$$

We already know the value of SSR. Since there are 2 variables, thus $k = 2$. To find out the SSE, we follow:

$$SSE = SST - SSR$$

$$SEE = 996.2 - 899.51$$

$$SEE = 96.69$$

So, the F test is:

$$F = \frac{899.51 / 2}{96.69 / (24 - 2 - 1)}$$

$$F = 97.77$$

The tabular F value for 2 and 21 degrees of freedom is 5.78 a 1% level. Thus the multiple linear regression is highly significant.

(d). The correlation coefficient is:

$$r = \sqrt{R^2}$$

$$r = \sqrt{0.90}$$

$$r = 0.94$$

There is a high degree of correlation between the dependent and independent variables.

3. An animal researcher is interested in the relationship between urea nitrogen and the energy balance and the phosphorous balance. The researcher has randomly selected 15 cows that have been fed rations that contain high and low levels of energy and phosphorous. The data for the experiment are given below.

Treatment Number	Y Urea N (mg/100 ml)	X_1 Energy Balance (k cal/day)	X_2 P Balance (g/day)
1	9.5	8.1	21.3
2	10.2	8.4	25.9
3	11.0	8.5	24.8
4	11.2	10.9	28.4
5	12.3	7.5	-2.7
6	10.8	7.8	22.2
7	10.8	11.3	19.0
8	11.1	-5.4	8.5
9	11.7	7.6	17.9
10	11.8	8.3	-9.9
11	10.9	9.6	-8.4
12	12.4	9.5	-10.1
13	12.6	10.5	12.4
14	10.9	-5.4	7.8
15	11.3	6.4	8.9
16	10.8	5.7	9.4

(a)   Estimate the least-square regression equation.
(b)   Compute the standard error of estimate.
(c)   Test the significance of the regression coefficients.
(d)   Compute the correlation coefficients.

Solution:

a.

Y	X_1	X_2	YX_1	X_1^2	X_1X_2	YX_2	X_2^2	Y^2
Urea N (mg/100 ml)	Energy Balance (k cal/day)	P Balance (g/day)						
9.5	8.1	21.3	76.95	65.61	172.53	202.35	453.69	90.25
10.2	8.4	25.9	85.68	70.56	217.56	264.18	670.81	104.04
11.0	8.5	24.8	93.50	72.25	210.80	272.80	615.04	121.00
11.2	10.9	28.4	122.08	118.81	309.56	318.08	806.56	125.44
12.3	7.5	-2.7	92.25	56.25	-20.25	-33.21	7.29	151.29
10.8	7.8	22.2	84.24	60.84	173.16	239.76	492.84	116.64
10.8	11.3	19.0	122.04	127.69	214.70	205.20	361.00	116.64
11.1	-5.4	8.5	-59.94	29.16	-45.90	94.35	72.25	123.21
11.7	7.6	17.9	88.92	57.76	136.04	209.43	320.41	136.89
11.8	8.3	-9.9	97.94	68.89	-82.17	-116.82	98.01	139.24
10.9	9.6	-8.4	104.64	92.16	-80.64	-91.56	70.56	118.81
12.4	9.5	-10.1	117.80	90.25	-95.95	-125.24	102.01	153.76
12.6	10.5	12.4	132.30	110.25	130.20	156.24	153.76	158.76
10.9	-5.4	7.8	-58.86	29.16	-42.12	85.02	60.84	118.81
11.3	6.4	8.9	72.32	40.96	56.96	100.57	79.21	127.69
10.8	5.7	9.4	61.56	32.49	53.58	101.52	88.36	116.64
179.3	109.3	175.4	1233.40	1123.10	1308.10	1882.70	4452.60	2019.10

$\bar{Y} = 11.21$ $\qquad$ $\bar{X}_1 = 6.83$ $\qquad$ $\bar{X}_2 = 10.96$

$$\Sigma y^2 = \Sigma Y^2 - \frac{(\Sigma Y)^2}{n} = 2019.1 - \frac{(179.3)^2}{16} = 9.82$$

$$\Sigma x_1^2 = \Sigma X_1^2 - \frac{(\Sigma X_1)^2}{n} = 1123.1 - \frac{(109.3)^2}{16} = 376.44$$

$$\Sigma x_2^2 = \Sigma X_2^2 - \frac{(\Sigma X_2)^2}{n} = 4452.6 - \frac{(175.4)^2}{16} = 2529.78$$

$$\Sigma x_1 y = \Sigma X_1 Y - \frac{(\Sigma Y)(\Sigma X_1)}{n} = 1233.4 - \frac{(179.3)(109.3)}{16} = 8.56$$

$$\Sigma x_2 y = \Sigma X_2 Y - \frac{(\Sigma Y)(\Sigma X_2)}{n} = 1882.7 - \frac{(179.3)(175.4)}{16} = -82.88$$

$$\Sigma x_1 x_2 = \Sigma X_1 X_2 - \frac{(\Sigma X_1)(\Sigma X_2)}{n} = 1308.1 - \frac{(109.3)(175.4)}{16} = 109.90$$

$$b_1 = \frac{\left(\Sigma x_2^2\right)\left(\Sigma x_1 y\right) - \left(\Sigma x_1 x_2\right)\left(\Sigma x_2 y\right)}{\left(\Sigma x_1^2\right)\left(\Sigma x_2^2\right) - \left(\Sigma x_1 x_2\right)^2}$$

$$b_2 = \frac{\left(\Sigma x_1^2\right)\left(\Sigma x_2 y\right) - \left(\Sigma x_1 x_2\right)\left(\Sigma x_1 y\right)}{\left(\Sigma x_1^2\right)\left(\Sigma x_2^2\right) - \left(\Sigma x_1 x_2\right)^2}$$

$$b_1 = \frac{(2529.78)(8.56) - (109.90)(-82.88)}{(376.44)(2529.78) - (109.90)^2} = 0.033$$

$$b_2 = \frac{(376.44)(-82.88) - (109.90)(8.56)}{(376.44)(2529.78) - (109.90)^2} = -0.034$$

$$a = \overline{Y} - b_1 \overline{X}_1 - b_2 \overline{X}_2$$

$$a = 11.21 - (0.033)(6.83) - (0.034)(10.96)$$

$$a = 11.35$$

(a).   The estimated multiple linear regression is:

$$\hat{Y} = 11.35 + 0.033X_1 - 0.034X_2$$

(b).   To compute the standard error of estimate, we must first calculate SSE as shown below:

$$SSE = SST - SSR$$

$$SST = \Sigma y^2 = 9.82$$

$$SSR = b_1 \Sigma x_1 y + b_2 \Sigma x_2 y$$

$$SSR = (0.033)(8.56) + (-0.034)(-82.88)$$

$$SSR = 3.10$$

Thus the SSE is computed as:

$$SSE = 9.82 - 3.10 = 6.72$$

To compute the standard error of estimate, we use the value of SSE in the following equation.

$$S_{y.12} = \sqrt{\frac{SSE}{n-k}}$$

$$S_{y.12} = \sqrt{\frac{6.72}{14}}$$

$$S_{y.12} = 0.69$$

(c). To test the significance of the coefficient of determination, we perform the F test as follows:

$$F = \frac{SSR / k}{SSE / (n - k - 1)}$$

$$F = \frac{3.10 / 2}{6.72 / (16 - 2 - 1)}$$

$$F = 2.98$$

The tabular F value for 2 and 13 degrees of freedom is 3.80 a 5% level. Thus the multiple linear regression is not significant. Caution must be taken in using the regression estimates.

(d). To compute the correlation coefficient we first compute the coefficient of determination.

$$R^2 = \frac{SSR}{SST}$$

$$R^2 = \frac{3.10}{9.82} = 0.32$$

$$r = \sqrt{R^2}$$

$$r = \sqrt{0.32}$$

$$r = 0.56$$

It appears that there is a weak correlation between the multiple linear regression variables.

4. An agricultural engineer wishes to determine the relationship between the monthly electrical usage in a greenhouse and the size of the greenhouse. He believes that the relationship is best exemplified by the following model:

$$Y = a + b_1 X_1 + b_2 X_2^2 + \varepsilon$$

Given the following data

Monthly electrical usage (kw/h)	Greenhouse size (ft^2)
2000	2800
2225	2900
2540	3100
2678	3200
2700	3250
2890	3300
2980	3400
3000	3500
3220	3550
3454	3600
3590	3650
3760	3850
4000	3900
4325	4000
4550	5050

(a) Estimate the multiple regression equation. (Use a computer to solve this problem.)

(b) Is the overall model useful in this investigation?

(c) Compute the correlation coefficients.

(d) Test the overall significance of the computed coefficients at $\alpha = .01$.

Solution:

a. To avoid working with large numbers in the computations, we have divided each data point by 100. The results are shown the in the table.

Y Monthly Electrical Use (kw/h)	X_1 Green House size (ft^2)	X_2	YX_1	X_1^2	X_1X_2	YX_2	X_2^2	Y^2
2.00	2.80	7.84	5.60	7.84	21.95	15.68	61.47	4.00
2.23	2.90	8.41	6.45	8.41	24.39	18.71	70.73	4.95
2.54	3.10	9.61	7.87	9.61	29.79	24.41	92.35	6.45
2.68	3.20	10.24	8.57	10.24	32.77	27.42	104.90	7.17
2.70	3.25	10.56	8.78	10.56	34.33	28.52	111.57	7.29
2.89	3.30	10.89	9.54	10.89	35.94	31.47	118.59	8.35
2.98	3.40	11.56	10.13	11.56	39.30	34.45	133.63	8.88
3.00	3.50	12.25	10.50	12.25	42.88	36.75	150.06	9.00
3.22	3.55	12.60	11.43	12.60	44.74	40.58	158.82	10.37
3.45	3.60	12.96	12.43	12.96	46.66	44.76	167.96	11.93
3.59	3.65	13.32	13.10	13.32	48.63	47.83	177.49	12.89
3.76	3.85	14.82	14.48	14.82	57.07	55.73	219.71	14.14
4.00	3.90	15.21	15.60	15.21	59.32	60.84	231.34	16.00
4.33	4.00	16.00	17.30	16.00	64.00	69.20	256.00	18.71
4.55	5.05	25.50	22.98	25.50	128.80	116.00	650.38	20.70
47.90	53.05	191.80	174.80	192.00	711.00	652.00	2705.00	160.80

$$\bar{Y} = 3.19 \qquad\qquad \bar{X}_1 = 3.54 \qquad\qquad \bar{X}_2 = 12.80$$

$$\Sigma y^2 = \Sigma Y^2 - \frac{(\Sigma Y)^2}{n} = 160.8 - \frac{(47.9)^2}{15} = 7.84$$

$$\Sigma x_1^2 = \Sigma X_1^2 - \frac{(\Sigma X_1)^2}{n} = 192 - \frac{(53.05)^2}{15} = 4.38$$

$$\Sigma x_2^2 = \Sigma X_2^2 - \frac{(\Sigma X_2)^2}{n} = 2705 - \frac{(191.8)^2}{15} = 252.53$$

$$\Sigma x_1 y = \Sigma X_1 Y - \frac{(\Sigma Y)(\Sigma X_1)}{n} = 174.8 - \frac{(47.9)(53.05)}{15} = 5.39$$

$$\Sigma x_2 y = \Sigma X_2 Y - \frac{(\Sigma Y)(\Sigma X_2)}{n} = 652 - \frac{(47.9)(191.8)}{15} = 39.52$$

$$\Sigma x_1 x_2 = \Sigma X_1 X_2 - \frac{(\Sigma X_1)(\Sigma X_2)}{n} = 711 - \frac{(53.05)(109.8)}{15} = 322.67$$

$$b_1 = \frac{\left(\Sigma x_2^2\right)\left(\Sigma x_1 y\right) - \left(\Sigma x_1 x_2\right)\left(\Sigma x_2 y\right)}{\left(\Sigma x_1^2\right)\left(\Sigma x_2^2\right) - \left(\Sigma x_1 x_2\right)^2}$$

$$b_2 = \frac{\left(\Sigma x_1^2\right)\left(\Sigma x_2 y\right) - \left(\Sigma x_1 x_2\right)\left(\Sigma x_1 y\right)}{\left(\Sigma x_1^2\right)\left(\Sigma x_2^2\right) - \left(\Sigma x_1 x_2\right)^2}$$

$$b_1 = \frac{(252.53)(5.39) - (322.67)(39.52)}{(4.38)(252.53) - (322.67)^2} = 0.11$$

$$b_2 = \frac{(4.38)(39.52) - (322.67)(5.38)}{(4.38)(252.53) - (322.67)^2} = -0.015$$

$$a = \bar{Y} - b_1 \bar{X}_1 - b_2 \bar{X}_2$$

$$a = 3.19 - (0.11)(3.54) - (-0.015)(12.8)$$

$$a = 2.99$$

(a). The estimated multiple linear regression is:

$$\hat{Y} = 2.99 + 0.11X_1 - 0.015X_2$$

(b). To determine whether the model is useful, we compute the coefficient of determination as

$$R^2 = \frac{SSR}{SST}$$

$$SST = \Sigma y^2 = 7.84$$

$$SSR = b_1 \Sigma x_1 y + b_2 \Sigma x_2 y$$

$$SSR = (0.11)(5.39) + (-0.015)(39.52)$$

$$SSR = 0.0001$$

$$R^2 = \frac{0.0001}{7.84} = 0.000012$$

The overall model is not useful as is implied by the coefficient of determination.

(c). The correlation coefficient is:

$$r = \sqrt{R^2}$$

$$r = \sqrt{0.000012}$$

$$r = 0.0035$$

As the computed correlation coefficient indicates, there is very small correlation between the variables.

(d). To test the significance of the model, we perform the F test as follows:

$$F = \frac{SSR / k}{SSE / (n - k - 1)}$$

$$SSR = 0.0001$$

$$SST = \Sigma y^2 = 7.84$$

Now that we know the SST and SSR, we can compute the SSE as follows:

$$SSE = SST - SSR$$

$$SSE = 7.84 - 0.0001 = 7.83$$

$$F = \frac{0.0001 / 2}{7.83 / (15 - 2 - 1)}$$

$$F = 0.00007$$

As the test shows, the results are highly insignificant.

5. Suppose that in the previous example, the agricultural engineer hypothesized that there is a strong relationship between the monthly electrical usage in the greenhouse and the size of the greenhouse, as well as the average monthly temperature. Data gathered are shown below.

Y Monthly electrical usage (kw/h)	X_1 Greenhouse size (ft^2)	X_2 Average monthly temperature (°F)
2000	2800	68
2225	2900	70
2540	3100	78
2678	3200	80
2700	3250	82
2890	3300	84
2980	3400	86
3000	3500	75
3220	3550	75
3454	3600	78
3590	3650	72
3760	3850	70
4000	3900	69
4325	4000	65
4550	5050	62

(a)   Compute the estimated regression equation. (Use a computer to solve this problem.)

(b)   Test the overall significance of the regression model when $\alpha = .05$.

Solution:

a. To avoid working with large numbers in the computations, we have divided each data point by 100. The results are shown the in the table.

Y	X_1	X_2	YX_1	X_1^2	X_1X_2	YX_2	X_2^2	Y^2
20.00	28.0	0.68	560.00	784.00	19.04	13.60	0.46	400.0
22.25	29.0	0.70	645.25	841.00	20.30	15.58	0.49	495.1
25.40	31.0	0.78	787.40	961.00	24.18	19.81	0.61	645.2
26.78	32.0	0.80	856.96	1024.00	25.60	21.42	0.64	717.2
27.00	32.5	0.82	877.50	1056.25	26.65	22.14	0.67	729.0
28.90	33.0	0.84	953.70	1089.00	27.72	24.28	0.71	835.2
29.80	34.0	0.86	1013.20	1156.00	29.24	25.63	0.74	888.0
30.00	35.0	0.75	1050.00	1225.00	26.25	22.50	0.56	900.0
32.20	35.5	0.75	1143.10	1260.25	26.63	24.15	0.56	1037.0
34.54	36.0	0.78	1243.40	1296.00	28.08	26.94	0.61	1193.0
35.90	36.5	0.72	1310.40	1332.25	26.28	25.85	0.52	1289.0
37.60	38.5	0.70	1447.60	1482.25	26.95	26.32	0.49	1414.0
40.00	39.0	0.69	1560.00	1521.00	26.91	27.60	0.48	1600.0
43.25	40.0	0.65	1730.00	1600.00	26.00	28.11	0.42	1871.0
45.50	50.5	0.62	2297.80	2550.25	31.31	28.21	0.38	2070.0
479.10	530.5	11.10	17476.00	19178.00	391.10	352.10	8.34	16083.0

$\bar{Y} = 31.94$ $\qquad \bar{X_1} = 35.37$ $\qquad \bar{X_2} = 0.74$

$$\Sigma y^2 = \Sigma Y^2 - \frac{(\Sigma Y)^2}{n} = 16083 - \frac{(479.10)^2}{15} = 780.5$$

$$\Sigma x_1^2 = \Sigma X_1^2 - \frac{(\Sigma X_1)^2}{n} = 19178 - \frac{(530.5)^2}{15} = 415.98$$

$$\Sigma x_2^2 = \Sigma X_2^2 - \frac{(\Sigma X_2)^2}{n} = 8.343 - \frac{(11.1)^2}{15} = 0.129$$

$$\Sigma x_1 y = \Sigma X_1 Y - \frac{(\Sigma Y)(\Sigma X_1)}{n} = 17476 - \frac{(479.1)(530.5)}{15} = 531.83$$

$$\Sigma x_2 y = \Sigma X_2 Y - \frac{(\Sigma Y)(\Sigma X_2)}{n} = 352.1 - \frac{(479.1)(11.1)}{15} = -2.43$$

$$\Sigma x_1 x_2 = \Sigma X_1 X_2 - \frac{(\Sigma X_1)(\Sigma X_2)}{n} = 391.1 - \frac{(530.1)(11.1)}{15} = -1.17$$

$$b_1 = \frac{\left(\Sigma x_2^2\right)\left(\Sigma x_1 y\right) - \left(\Sigma x_1 x_2\right)\left(\Sigma x_2 y\right)}{\left(\Sigma x_1^2\right)\left(\Sigma x_2^2\right) - \left(\Sigma x_1 x_2\right)^2}$$

$$b_2 = \frac{\left(\Sigma x_1^2\right)\left(\Sigma x_2 y\right) - \left(\Sigma x_1 x_2\right)\left(\Sigma x_1 y\right)}{\left(\Sigma x_1^2\right)\left(\Sigma x_2^2\right) - \left(\Sigma x_1 x_2\right)^2}$$

$$b_1 = \frac{(0.129)(531.83) - (-1.17)(-2.43)}{(415.98)(0.129) - (-1.17)^2} = 1.26$$

$$b_2 = \frac{(415.98)(-2.43) - (-1.17)(531.83)}{(415.98)(0.129) - (-1.17)^2} = -7.43$$

$$a = \overline{Y} - b_1 \overline{X}_1 - b_2 \overline{X}_2$$

$$a = 31.94 - (1.26)(35.37) - (-7.43)(0.74)$$

$$a = -7.13$$

(a).   The estimated multiple linear regression is:

$$\hat{Y} = -7.13 + 1.26X_1 - 7.43X_2$$

(b).   To test the significance of the model, we perform the F test as follows:

$$F = \frac{SSR / k}{SSE / (n - k - 1)}$$

$$SSR = b_1 \Sigma x_1 y + b_2 \Sigma x_2 y$$

$$SSR = (1.26)(531.83) + (-7.43)(-2.43)$$

$$SSR = 688.16$$

$$SST = \Sigma y^2 = 780.5$$

$$SSE = SST - SSR = 780.5 - 688.16 = 92.34$$

Now we perform the F test:

$$F = \frac{688.16 / 2}{92.34 / (15 - 2 - 1)} = 44.74$$

The F test results show that the estimates of the regression are significant at the 5% level.

6. An equine researcher has postulated that there is a nonlinear relationship between the life span of Arabian horses and the gestation period. The researcher believes that the following model best exemplifies the relationship:

$$Y = a + b_1 X_1 + b_2 X_2^2 + \varepsilon$$

Life Span (years)	Gestation Period (days)
15.2	210
17.8	230
18.2	240
20.0	265
22.1	285
23.4	370
20.5	345
22.2	340
21.5	385
19.6	295
18.8	279
20.4	305
22.3	315
20.0	395
18.4	290
15.5	300

(a) Estimate the multiple regression equation. (Use a computer to solve this problem.)

(b) Interpret the meaning of the coefficient of multiple determination.

(c) Are the regression coefficient significant at the $\alpha = .01$?

(d) Test the overall significance of the model at $\alpha = .01$.

Polynomial Regression X_1: X1 Y_1: Column 1

DF:	R:	R-squared:	Adj. R-squared:	Std. Error:
15	.691	.478	.397	1.829

Analysis of Variance Table

Source	DF:	Sum Squares:	Mean Square:	F-test:
REGRESSION	2	39.752	19.876	5.942
RESIDUAL	13	43.487	3.345	p = .0147
TOTAL	15	83.239		

Residual Information Table

SS[e(i)-e(i-1)]:	e ≥ 0:	e < 0:	DW test:
39.102	9	7	.899

1

Polynomial Regression X_1: X1 Y_1: Column 1

Beta Coefficient Table

Parameter:	Value:	Std. Err.:	Std. Value:	t-Value:	Probability:
INTERCEPT	-5.604				
x	.141	.094	3.23	1.51	.1549
x^2	-1.849E-4	1.524E-4	-2.595	1.213	.2467

2

Polynomial Regression X_1: X1 Y_1: Column 1

Confidence Intervals and Partial F Table

Parameter:	95% Lower:	95% Upper:	90% Lower:	90% Upper:	Partial F:
INTERCEPT					
x	-.061	.344	-.024	.307	2.281
x^2	-.001	1.444E-4	-4.548E-4	8.503E-5	1.472

3

Solution:

The computer printouts are given below.

Polynomial Regression X_2: X2 Y_1: Column 1

DF:	R:	R-squared:	Adj. R-squared:	Std. Error:
15	.694	.482	.402	1.821

Analysis of Variance Table

Source	DF:	Sum Squares:	Mean Square:	F-test:
REGRESSION	2	40.123	20.062	6.049
RESIDUAL	13	43.116	3.317	p = .0139
TOTAL	15	83.239		

Residual Information Table

SS[e(i)-e(i-1)]:	e ≥ 0:	e < 0:	DW test:
38.174	9	7	.885

4

Polynomial Regression X_2: X2 Y_1: Column 1

Beta Coefficient Table

Parameter:	Value:	Std. Err.:	Std. Value:	t-Value:	Probability:
INTERCEPT	9.566				
x	1.767E-4	8.647E-5	2.481	2.044	.0618
x^2	-6.556E-10	4.221E-10	-1.885	1.553	.1444

5

Polynomial Regression X_2: X2 Y_1: Column 1

Confidence Intervals and Partial F Table

Parameter:	95% Lower:	95% Upper:	90% Lower:	90% Upper:	Partial F:
INTERCEPT					
x	-1.010E-5	3.635E-4	2.358E-5	3.299E-4	4.177
x^2	-1.568E-9	2.564E-10	-1.403E-9	9.201E-11	2.412

6